Mediterranean Great White Sharks

Mediterranean Great White Sharks

*A Comprehensive Study
Including All Recorded Sightings*

ALESSANDRO DE MADDALENA *and*
WALTER HEIM

Foreword by MARIE LEVINE

McFarland & Company, Inc., Publishers
Jefferson, North Carolina, and London

ALSO OF INTEREST

Great White Sharks in United States Museums
(by Alessandro De Maddalena and Walter Heim; McFarland 2009)

All drawings © 2012 Alessandro De Maddalena

LIBRARY OF CONGRESS CATALOGUING-IN-PUBLICATION DATA

De Maddalena, Alessandro, 1970–
Mediterranean great white sharks : a comprehensive study including all recorded sightings / Alessandro De Maddalena and Walter Heim ; foreword by Marie Levine.
p.   cm.
Includes bibliographical references and index.

ISBN 978-0-7864-5889-9
softcover : acid free paper ∞

1. White shark — Mediterranean Sea — Case studies.
I. Heim, Walter (Walter D.)   II. Title.
QL638.95.L3D4137   2012   597.3'3091638 — dc23   2011053201

British Library cataloguing data are available

© 2012 Alessandro De Maddalena and Walter Heim. All rights reserved

No part of this book may be reproduced or transmitted in any form or by any means, electronic or mechanical, including photocopying or recording, or by any information storage and retrieval system, without permission in writing from the publisher.

Manufactured in the United States of America

On the cover a young female white shark caught on April, 20, 2006, off Aras Dizra, Tunisia (photograph by Samuel P. Iglésias / Muséum National d'Histoire Naturelle de Paris); background images © 2012 Shutterstock

McFarland & Company, Inc., Publishers
Box 611, Jefferson, North Carolina 28640
www.mcfarlandpub.com

To my wonderful wife, Alessandra, and my sweet son,
Antonio, with deep love. — Alessandro

To my daughters, Rachel and Jessica, who as young girls spent
many days chasing sharks with their dad. — Walter

Contents

Acknowledgments v
Foreword by Marie Levine 1
Preface 3

1 — The Great White Shark 5
2 — Records of Great White Sharks from the Mediterranean Sea 16

    Spain 16
      Continental Spain 16
      Islas Baleares 18
    France 22
      Continental France 22
      Corsica 31
    Monaco 32
    Italy 32
      Ligurian Coast 33
      Sardinia 38
      Tyrrhenian Coast 42
      Sicily 54
      Ionian Coast 67
      Adriatic Coast 68
    Slovenia 77
    Croatia 78
    Bosnia and Herzegovina 86
    Montenegro 86
    Albania 86
    Greece 86
    Turkey 88
    Syria 92
    Cyprus 92
    Lebanon 92
    Israel 92
    Egypt 92
    Libya 93
    Malta 94
    Tunisia 99
    Algeria 105
    Morocco 106
    Unknown Locations 106
      Adriatic Sea 106
      Mediterranean Sea 107
      Specimens Without a Capture Location 107

3 — Summary Tables 113
4 — Analysis of the Presence of Great White Sharks in the Mediterranean Sea 175

    Common Names Used in the Mediterranean Area 175
    Similar Species Present in the Study Area 175
    Maximum Size 176
    Average Size 179
    Color 179
    Reproduction 179
    Distribution 182
    Identification 184
    Habitat, Movements and Seasonality 185
    Diet 188
    Predatory Tactics 192
    Scavenging 194

Predators and Parasites 196
Mutualism 198
Attacks on Humans 198
Fisheries 203
Interactions with Tuna
  Farming 208
Utilization 211
Strandings 213

Abundance 214
Conservation 216
White Shark Materials Preserved in
  Museums 217
Photographic and Filmed
  Documentation 218
Reporting Specimens of Great White
  Sharks from the Mediterranean Sea 220

Appendix: Mediterranean Great White Shark Report Form   223
Bibliography   225
Index   237

Acknowledgments

We must pay special homage to Beverly Heim (San Diego, California) who took the time to read and edit the entire manuscript. Many thanks to respected colleague, shark expert, and dear friend Marie Levine for taking her precious time to read the manuscript and write the foreword.

We thank the following people, as well as the institutions where they work, for freely sharing their observations and for their assistance in assembling material for this book: Daniel Abed-Navandi (Haus des Meeres, Aqua Terra Zoo, Wien, Austria), Peter Adamik (Vlastivědné muzeum v Olomouci, Olomouc, Czech Republic), Luigi Alberotanza (Consiglio Nazionale delle Ricerche, Istituto per lo Studio delle Dinamica delle Grandi Masse, Venezia, Italy), Nicola Allegri (Italy), Luca Altichieri (Museo di Geologia e Paleontologia, Università di Padova, Padova, Italy), Gérard Altman (Association Nationale de Moniteur de Plongée, Antibes, France), Alain Alziari (Bataillon de Marins-Pompiers de Marseille, Marseille, France), Scot Anderson (Inverness, California, USA), Riccardo Andreoli (Venezia, Italy), Ivano Ansaloni (Museo di Storia Naturale e della Strumentazione Scientifica, Modena, Italy), Jean Attard (Apnéa, Toulouse, France), Nathaniel Attard (*In-Nazzjon*, Media Link Communications, Malta), Gerhard Aubrecht (Biologiezentrum, Oberoesterreichische Landesmuseen, Linz/Dornach, Austria), Vittorio Ballerini (Cetaria Diving Center, Scopello, Italy), Emilio Balletto (Dipartimento di Biologia Animale e dell'Uomo, Università degli studi di Torino, Torino, Italy), Enrico Banfi (Museo Civico di Storia Naturale, Milano, Italy), Manolis Bardanis (Naxos Diving, Apeiranthos — Naxos, Greece), Alessandro Barlettani (Italy), Joan Barrull (Museu de Zoologia, Barcelona, Spain), Àlex Bartolí (CRAM — Fundació per a la Conservació i Recuperació d'Animals Marins, Premiá de Mar, Spain), Peter Bartsch (Museum für Naturkunde, Humboldt-Universitaet zu Berlin, Berlin, Germany), Emilio Balletto (Dipartimento di Zoologia, Università di Torino, Torino, Italy), Saulo Bambi (Museo di Storia Naturale dell'Università di Firenze, Sezione di Zoologia "La Specola," Firenze, Italy), Roberto Basso (Museo civico di Storia Natural di Jesolo, Jesolo, Italy), Edward R. Battisti (Panathlon Club, Brescia), David C. Bernvi (Caracal Publishing, Gothenburg, Sweden), Miguel Berrios (NOAA Fisheries, Pacific Islands Region Observer Program, Honolulu, Hawaii, USA), Didier Berthet (Centre de Conservation et d'Etude des Collection, Musée des Confluences, Lyon, France), Christian Bertok (Trieste, Italy), Vinicio Biagi (alla memoria), Ezio Bocedi (Carrara, Italy), Roberto Boracci (Italy), Giovanni Bosco (Italy), Enrico Borgo (Museo Civico di Storia Naturale "Giacomo Doria," Genova, Italy), Mario Bozzi Editore (Genova, Italy), Mohamed Nejmeddine Bradaï (Laboratoire Biodiversité et Biotechnologie Marines, Institut National des Sciences et Technologies de la Mer, Sfax, Tunisie), Pierre Brocchi (Aquanaude, Nice, France), Andrea Bryk (Haus der Natur, Salzburg, Austria), John Clay Bruner (Department of Biological Sciences, University of Alberta, Edmonton, Alberta, Canada), Michèle Bruni (Musée Océanographique, Monaco-Ville, Principauté de Monaco), Alex "The Sharkman" Buttigieg (Sharkman's World Organization, San Gwann, Malta), Philippe Candegabe (Muséum d'histoire naturelle de Grenoble, Grenoble, France), Ernesto Capanna (Facoltà di Scienze Matematiche, Fisiche e Naturali, Università di Roma "La Sapienza," Roma, Italy), Christian Capapé (Laboratoire d'Ichtyologie, Université Montpellier II, Sciences et Techniques du Languedoc, Montpellier, France), Henri Cappetta (Institut des Sciences de l'Evolution, Université de Montpellier II — Sciences et Techniques du Languedoc, Montpellier, France), Stefano Carletti (Italy), Baldassare Carollo (Alcamo, Italy), Stefano Catalani (Senigallia, Italy), Gioacchino Cataldo (Tonnara di Favignana, Favignana, Italy), Luigi Cavaleri (Consiglio Nazionale delle Ricerche, Istituto per lo Studio delle Dinamica delle Grandi Masse, Venezia, Italy), Antonio Celona (Aquastudio Research

Acknowledgments

Institute, Messina, Italy), Silvia Chicchi (Musei Civici, Reggio Emilia, Italy), Giuliano Chiocca (Italy), Franco Cigala Fulgosi (Dipartimento di Scienze della Terra, Università di Parma, Parma, Italy), Salvatore Cilona (Italy), Geremy Cliff (Natal Sharks Board, Umhlanga Rocks, South Africa), Vito Cofrancesco (Italy), Ralph Collier (Shark Research Committee, Van Nuys, California, USA), Stefano Alberto Colombo (Zara Point, Milano, Italy), Giorgio Colombo (Busto Arsizio, Italy), Marco Colombo (Busto Arsizio, Italy), Giorgia Comparetto (Necton Marine Research Society, Catania, Italy), Mauro Cottiglia (Istituto di Zoologia, Università degli Studi di Cagliari, Cagliari, Italy), Evelyne Crégut (Muséum Requien, Avignon, France), Oliver Crimmen (British Museum of Natural History, London, England, United Kingdom), Gianluca Cugini (Mediterranean Shark Research Group, Pescara, Italy), Mohamed Dahmouni (Aquasea, Mahdia, Tunisia), Emiliano D'Andrea (Società di Ricerca Necton, Ganzirri, Italy), Michael Darmanin (Malta Centre for Fisheries Sciences, Department of Fisheries and Aquaculture, Fort San Lucjan, Marsaxlokk, Malta), Leonardo Leone De Castris (Brindisi, Italy), Andrea Del Coco (Lecce, Italy), Gianfranco Della Rovere (Milano, Italy), Andrea dall'Asta (Museo Civico di Storia Naturale di Trieste, Trieste, Italy), Luigi De Giosa (Alessano, Italy), Isabella De Maddalena (Milano, Italy), Gregorio De Metrio (Dipartimento di Sanità e Benessere degli Animali, Università degli Studi di Bari, Valenzano, Bari), Pascal Deynat (Laboratoire d'Ichtyologie Générale et Appliquée, Museum National d'Histoire Naturelle de Paris, Paris, France), Sergio Dolce (Museo Civico di Storia Naturale, Trieste, Italy), Antonino Donato (Ganzirri, Italy), Giuliano Doria (Museo Civico di Storia Naturale "Giacomo Doria," Genova, Italy), Branko Dragicevic (Laboratory of Ichthyology and Coastal Fishery, Institute of Oceanography and Fisheries, Split, Croatia), Clinton Duffy (Marine Conservation Unit, Department of Conservation, Auckland, New Zealand), Jakov Dulcic (Laboratory of Ichthyology and Coastal Fishery, Institute of Oceanography and Fisheries, Split, Croatia), Richard Ellis (American Museum of Natural History, New York, USA), Jean-Claude Eugéne (Club Marseille Sports, Marseille, France), Juan Manuel Ezcurra (Monterey Bay Aquarium, Monterey Bay, California, USA), Riccardo Fanelli (Italy), Gino Felicioni (Italy), Jordi Ferré (TV Tossa, Tossa de Mar, Spain), Fausto Fioretti (Italy), Pedro Fominaya (Spain), Donatella Foddai (Dipartimento di Biologia, Università degli Studi di Padova, Padova, Italy), Nicola Franzese (Museo Regionale di Scienze Naturali, Torino, Italy), Txema Galaz (Tuna Farms of Mediterraneo, San Javier, Murcia, Spain), Gildo Gavanelli (Progetto *Mola mola*, Imola, Italy), Elena Gavetti (Museo Regionale di Scienze Naturali, Torino, Italy), Christian Gelpi (Club Nausicaa, Nice, France), Olivier Gerriet (Muséum d'Histoire Naturelle de Nice, Nice, France), Roberto Gioia (Roma, Italy), Vincenzo Giudice (Italy), Olivier Glaizot (Musée cantonal de Zoologie, Lausanne, Switzerland), Gérard Gory (Muséum d'histoire naturelle de Nîmes, Nîmes, France), Luciana Grambo (Italy), Stéphane Granzotto (Paladru, France), Kurt Grossenbacher (Naturhistorisches Museum, Bern, Switzerland), Giuseppe Guarrasi (Favignana, Italy), John Gullaumier (Malta), Eric G. Haenni (Crossroads Environmental Consultants, Palm City, Florida, USA), Mohamed Hamdine (Faculté Centrale d'Alger, Institut des Sciences de la Mer et de l'Amenagement du Littoral, Alger, Algeria), Thomas Hammann (Pohlheim, Germany), Adem Hamzic (University of Sarajevo, Faculty of Science, Ichthyology and Fishing Center, Sarajevo, Bosnie-Herzégovine), Bill Heim (Alexandria, Virginia, USA), Walter Heim (San Diego, California, USA), Ernst Hofinger (Hofinger Tier-Präparationen, Steyrermühl, Austria), Samuel P. Iglésias, Muséum national d'Histoire naturelle, Station de Biologie Marine de Concarneau, Concarneau, France), Ruggero Ilgrande (Milano, Italy), Hakan Kabasakal (Ichthyological Research Society, Istanbul, Turkey), Zoran Kljajic (Institute of Marine Biology, Kotor, Montenegro), Drazen Kotrošan (Zemaljski muzej Bosne i Hercegovine, Sarajevo, Bosnia and Herzergovina), Marcelo Kovačić (Prirodoslovni muzej Rijeka, Rijeka, Croatia), Michel Krafft (Musée cantonal de Zoologie, Lausanne, Switzerland), Friedhelm Krupp (Senckenberg Forschungsinstitut und Naturmuseum, Frankfurt a. M., Germany), Boris Krystufek (Prirodoslovni muzej Slovenije, Ljubljana, Slovenija), Philippe Laulhe (Strasbourg, France), Georges Lenglet (Koninklijk Belgisch Instituut voor Natuurwetenschappen, Brussels, Belgium), Dino Levi (Istituto di Ricerca sulle Risorse Marine e l'Ambiente, Mazara del Vallo, Italy), Marie Levine

Acknowledgments

(Global Shark Attack File, Princeton, New Jersey, USA), Lovrenc Lipej (Marine Biological Station, National Institute of Biology, Piran, Slovenia), Jeff Liston (Hunterian Museum, Glasgow, Scotland, United Kingdom), Robert Lončarić (Department of Geography, University of Zadar, Zadar, Croatia), Giacomo Longhi (Italy), Walid Maamouri (Tunisia), Nicola Maio (Museo Zoologico, Napoli, Italy), Tore Manca (Italy), Vincent Maliet (Collectivité Territoriale de Corsica, Ajaccio, France), Renata Mangano (DELPHIS Aeolian Dolphin Center, Salina, Italy), Riccardo Manni (Dipartimento di Scienze della Terra, Università degli Studi "La Sapienza," Roma, Italy), Gianni Marangoni (Museo Civico di Zoologia, Giardino Zoologico, Roma, Italy), Domenico Marcianò (Italy), Mario Marconi (Museo di Scienze Naturali, Università degli Studi di Camerino, Camerino, Italy), Renato Mariani-Costantini (Department of Oncology and Experimental Medicine, "G. d'Annunzio" University, Chieti-Pescara, Italy), R. Aidan Martin (ReefQuest Centre for Shark Research, Vancouver, British Columbia, Canada), Martín Martínez Lorca (Spain), Adelaide Mastandrea (Istituto di Paleontologia, Università degli Studi di Modena, Modena, Italy), Isabel Mate (Museu de Zoologia, Barcelona, Spain), Stéphane Mattei (Studio B, Sagone, France), Elvio Mazzagufo (Italy), Glauco Micheli (Italy), Vincenzo Milanesi (Università degli Studi di Padova, Padova, Italy), Lorenzo Millan (Tuna Farms of Mediterraneo, San Javier, Murcia, Spain), Alessandro Minelli (Dipartimento di Biologia, Università degli Studi di Padova, Padova, Italy), Daniela Minelli (Museo di Anatomia Comparata, Bologna, Italy), Nitto Mineo (Favignana, Italy), Roger Miniconi (Conseil scientifique régional du patrimoine naturel, Corsica, France), Luca Mizzan (Museo Civico di Storia Naturale Fontego dei Turchi, Venezia, Italy), Angelo Mojetta (Acquario e Civica Stazione Idrobiologica, Milano, Italy), Francesc Xavier Viñals Moncusi (Reus, Spain), Neil Montoya (USA), Jiří Moravec (Národní Muzeum, Praha, Czech Republic), Nicolas Morel (Musée Vert — Musée d'Histoire Naturelle du Mans, Le Mans, France), Alessandro Morescalchi (Istituto di Anatomia Comparata, Università degli Studi di Genova, Genova, Italy), Gabriel Morey (Direcció General de Pesca, Conselleria d'Agricultura i Pesca — Govern de les Illes Balears, Palma de Majora, Spain), Famiglia Murru (Capo Testa, Italy), Ulisse G. Murru (Capo Testa, Italy), Lisa J. Natanson (National Marine Fisheries Service, Narragansett, Rhode Island, USA), Marco Maria Navoni (Biblioteca Ambrosiana, Milano, Italy), Paola Nicolosi (Museo di Zoologia, Università degli Studi di Padova, Padova, Italy), Jørgen Nielsen (Zoologisk Museum, University of Copenhagen, København, Denmark), Guy Oliver (Laboratoire de Biologie Physico-Chimique, Université de Perpignan, Perpignan, France), Maurizio Omodei (Scopello, Italy), Lidia Orsi Relini (DIPTERIS, Università di Genova, Genova, Italy), Richard Peirce (Richard Peirce Shark Conservation, Bude, Cornwall, UK), Carlo Pesarini (Museo Civico di Storia Naturale, Milano, Italy), *Pesca in Mare* (Italy), Luigi Piscitelli (Società Ittiologica Italiana, Milano, Italy), Angelo Pistorio (Italy), Jürgen Plass (Biologiezentrum, Oberoesterreichische Landesmuseen, Linz/Dornach, Austria), Michela Podestà (Museo Civico di Storia Naturale, Milano, Italy), Marta Poggesi (Museo di Storia Naturale dell'Università di Firenze — Sezione di Zoologia "La Specola," Firenze, Italy), Roberto Poggi (Museo Civico di Storia Naturale "Giacomo Doria," Genova, Italy), Cesare Polidori (Italy), Vittorio Pomante (Italy), Antonella Preti (National Marine Fisheries Service, Southwest Fisheries Science Center, La Jolla, California, USA), Peter Psomadakis (Museo Zoologico, Napoli, Italy), Gianfranco Purpura (Dipartimento di Storia del Diritto, Università di Palermo, Palermo, Italy), Antonino Rallo (Favignana, Italy), Ninni Ravazza (Cosedimare, Trapani, Italy), Alberto Luca Recchi (RAL Gruppo srl, Roma, Italy), Giulio Relini (DIPTERIS, Università di Genova, Genova, Italy), Anne-Lyse Révelart (Praha, Czech Republic), Danilo Rezzolla (Mediterranean Shark Research Group, Milano, Italy), Klaus Riediger (Austria), Rhian Rowson (Bristol Museum and Art Gallery, Bristol, England, United Kingdom), Alessandro Russo (Ufficio Circondariale Marittimo, Piombino, Italy), Mauro Salomone (Società Capitani e Macchinisti Navali Camogli, Camogli, Italy), Radek Šanda (Národní Muzeum, Praha, Czech Republic), Paolo Santi (Italy), Maurizio Sarà (Museo di Zoologia Doderlein, Palermo, Italy), Michael C. Scholl (White Shark Trust, Saint-Légier, Switzerland), Teddi Sciurti (Italy), Fabrizio Serena (ARPAT-

Acknowledgments

AREAMARE, Livorno, Italy), Bernard Séret (Laboratoire d'Ichtyologie Générale et Appliquée, Museum National d'Histoire Naturelle de Paris, Paris, France), Paolo Sibille (Museo Regionale di Scienze Naturali, Torino, Italy), Giovanni Sigovini (AUSL 12 Veneziana, Mercato Ittico all'Ingrosso di Venezia, Venezia, Italy), Lobert Simičić (Rava, Croatia), Thomas Sohm (Inatura — Erlebnis Naturschau Dornbirn, Dornbirn, Austria), Alen Soldo (Laboratory of Ichthyology and Coastal Fishery, Institute of Oceanography and Fisheries, Split, Croatia), Pascale Soleil (Musée des beaux-arts et d'archéologie, Valence, France), Mommo Solina (Bonagia, Italy), D. Sorrenti (Ganzirri, Italy), Gerlando Spagnolo (Lamezia Terme, Italy), Salvatore Spataro (Favignana, Italy), Paolo Spinelli (Italy), Tiziano Storai (Museo Civico di Scienze Naturali della Valdinievole, Pescia, Italy), Jérôme Streng (France), Philippe Summonti (Centre d'Intéret de Quartier de l'Estaque, Marseille, France), Marco Tarantino (Firenze, Italy), Skip Theberge (NOAA Central Library, Silver Spring, Maryland, USA), Paolo Tongiorgi (Museo di Storia Naturale e della Strumentazione Scientifica, Modena, Italy), Sven Traenkner (Senckenberg Forschungsinstitut und Naturmuseum, Frankfurt a. M., Germany), Sam Trebilcock (Bristol Museum and Art Gallery, Bristol, England, United Kingdom), Géry Van Grevelynghe (Stella, Piton Saint-Leu, La Réunion, France), Stefano Vanni (Museo di Storia Naturale dell'Università di Firenze — Sezione di Zoologia "La Specola," Firenze, Italy), Paolo Virnicchi (Italy), Alessandro Vitturini (Italy), Cecilia Volpi (Museo di Storia Naturale dell'Università di Firenze — Sezione di Zoologia "La Specola," Firenze, Italy), Claude Wagner (Abyss Explorers, Marseille, France), Pierre-Henri Weber (Corsica Mare Osservazione, Ajaccio, France), Helmut Wellendorf (Naturhistorisches Museum Wien, Wien, Austria), Thomas Winkler (Tierpräparation Thomas Winkler, Trebus, Germany), Sabine Wintner (Natal Sharks Board, Umhlanga Rocks, South Africa), Alfred Xuereb (Malta), Alberto Zaccagni (Italy), Alberto Zanoli (Italy), Marco Zuffa (Museo Archeologico "Luigi Donini," Ozzano dell'Emilia, Italy), Marco Zuffi (Museo di Storia Naturale e del Territorio, Calci, Italy).

For their help, support and friendship, our sincere gratitude goes to Alessandra Baldi, Antonio De Maddalena, Pinuccia De Maddalena, Emilio De Maddalena, Eleonora De Maddalena, Elisabetta De Maddalena, Isabella De Maddalena, Andrea Del Coco, Sauro Baldi, the students of the BCM school, Francesco Guerrazzi, Gianfranco Della Rovere, Matteo Messa, Michele Masera, Antonella Preti, Massimo Albini, Gaspare Schillaci, Claudio Perotti, Giorgio Martello Panno, Alessandro De Marinis, Chris Fallows, Monique Fallows, Ralph S. Collier, Sean R. Van Sommeran, Brian May, Paul Rodgers, the Mediterranean Shark Research Group and the Italian Ichthyological Society.

Alessandro De Maddalena would also like to thank Jon Anderson, who has been a great source of inspiration with his music in the preparation of this work.

Foreword by Marie Levine

During the past forty years, interest in white sharks has grown enormously. Exciting discoveries by scientists throughout the world have contributed to a growing interest in these magnificent predators. Nevertheless, I have searched for such a book as this in libraries on three continents for more than three decades, but it had not yet been written. At long last it is here: A comprehensive, meticulously researched book on white sharks in the Mediterranean.

It is exceedingly difficult to retrieve archival information; records are destroyed, witnesses pass away and vital data may be lost forever. The scientific community will always be thankful to Dott. De Maddalena for gathering, collating and safeguarding the wealth of archival information contained in the Italian Great White Shark Data Bank. Those of us who study the white shark are especially indebted to the authors. The painstaking research and long years of study which Dott. De Maddalena, with the collaboration of Walter Heim, put into the writing of this book has made it a major resource on white sharks.

This book is a rich resource of information for historians, scientists, fishermen, and divers and non-divers alike. It is a book for everyone who wishes to learn more about white sharks.

Marie Levine is founder of the
Shark Research Institute, Princeton, New Jersey.

Preface

Alessandro De Maddalena

This book is a milestone in the research on white sharks. It is the first complete publication about the presence of the great white shark in the Mediterranean Sea. In this book, the data and results of a study that started in 1996 are finally presented in whole to the scientific community. In order to collect and analyze the available information about the great white sharks in the Mediterranean, I started research on the records of this large predator in those waters. These records were organized into a data bank, which was initially developed as my thesis for the completion of my studies in natural sciences at the University of Milan. However, the large amount of data obtained from this research, along with the interest in this project shown by many shark researchers, the general public and the media, convinced me to continue my work on it. At this point, the work was named the Italian Great White Shark Data Bank. Today, this data bank includes information on 596 records of great white sharks from the entire Mediterranean Sea, representing the most complete and comprehensive study ever performed on the great white sharks in that area.

The collaboration of many researchers, especially the members of the Mediterranean Shark Research Group, has been fundamental in the completion of this work. Another important source of data has been the numerous shark enthusiasts with whom I have been in communication throughout the years. But the most important contribution has come from occasional observers, including divers, seamen, commercial fishermen, sport fishermen and the coast guard who unexpectedly encountered a great white shark.

The data collected since the creation of the Italian Great White Shark Data Bank includes a substantial amount of information on size, distribution, habitat, behavior, reproduction, diet, fishery and attacks on humans. Over the years, parts of the findings of this study have been presented in a number of scientific articles, popular articles and four books. In these works, only specific aspects of the study have been discussed. Two of the four books are written in Italian: *Lo squalo bianco nei mari d'italia* (Ireco, 2002) and *Lo squalo bianco nel mare mediterraneo* (Rivista Marittima, 2010). One book was written in French: *Le Grand Requin blanc sur les côtes françaises* (Turtle Prod Éditions/Média Plongée, 2008). The fourth book was written in English and focuses on great white sharks in museums: *Great White Sharks Preserved in European Museums* (Cambridge Scholars, 2007). These publications have introduced marine biologists and the general public to the biology, ethology and ecology of the Mediterranean great white sharks.

After all these years of study on this subject, I felt it was now time to present the entire work completed to date in a single, exhaustive publication, including all data previously presented in a number of different publications, as well as a large amount of information that has remained unpublished. This book presents all the data collected in the Great White Shark Data Bank and is finally available to the shark specialists as well as to the general public. I hope the readers will find this publication interesting and useful and that it will stimulate further research in this field.

Walter Heim

The white shark is a magnificent creature. It haunted the Mediterranean Sea long before man dared to venture on the water. Imagine a time in the Mediterranean long before records. A fisherman paddles his small craft off the shore and encounters a six meter long white shark. Maybe the shark is curious or maybe it is aggressive. The sighting of this large massive shark must have petrified the poor fisherman. The data bank indicates that Mediterranean white sharks occasionally attack boats. So maybe this poor fisherman perishes as the shark capsizes the boat. More likely, the shark is curious and investigates the fisherman before moving on. Along with sea monsters, stories of monster sharks strike fear among people and these stories and fears pass down through the generations.

Centuries later, man has become the dominant predator, but he still fears the white shark. Huge sharks are caught and displayed to the public. In fact, as the records show, there was a time when man placed a bounty on the Mediterranean white sharks. Today, the Mediterranean white shark is still feared, but as more becomes known about it, we find that this huge predator does not stalk man but prefers other prey. Man is a curiosity and seldom a meal. The Italian Great White Shark Data Bank takes the reader through the history of great white shark encounters, from records as early as the Middle Ages involving knights in armor to the present such as white sharks in conflict with bluefin tuna farms. The data bank records encounters between white sharks and bathers, divers, boats, fishermen and others. In most cases, the white shark is the loser, as it is caught and killed. Today, Mediterranean white sharks are in decline and are a protected species. Hopefully, the population will rebound and be available for our progeny to respect and not fear. The history of white shark encounters in the Mediterranean is fascinating and we hope the reader will enjoy reading through this compilation of nearly 600 records of Mediterranean white sharks.

1

The Great White Shark

General Anatomical Characteristics of the Great White Shark

Great white sharks belong to the kingdom Animal, the phylum Chordata, the subphylum Vertebrata, the superclass Pisces, the class Chondrichthyes (cartilaginous fish), the subclass Elasmobranchii (sharks and rays), the superorder Selachimorpha (sharks), the order Lamniformes (mackerel sharks), the family Lamnidae, and the genus *Carcharodon*. The genus *Carcharodon* includes only one living species, *Carcharodon carcharias* (Linnaeus, 1758) — the great white shark. The Latin name of the great white shark genus is derived from the Greek *kàrkharos*, which means "serrated," and *odón*, which means "tooth," referring to the strongly serrated margins of the teeth. The earliest fossils in the genus *Carcharodon* have been found in the Paleocene (55.8 to 65.5 million years ago). *Carcharodon carcharias* did not appear until the Late Miocene (11.6 to 5.3 million years ago) (Applegate and Espinosa-Arrubarrena, 1996).

Great white sharks are easy to recognize. They are typically large in size, ranging in length from 120 cm to at least 700 cm. The great white shark is a pelagic species with a massive, spindle-shaped body. The snout is large, conical and pointed. The mouth is wide and the lower teeth are prominent and visible even when the mouth is closed. The eyes are relatively small, dark, and circular in shape, lacking a nictitating membrane. The nostrils are relatively small, located lateroventrally on the snout. The nostrils are partially covered by a nasal flap that separates sea water flowing into the nostril from water flowing out. There are very small spiracles located on the sides of the head, posterior to the eyes and anterior to the gill slits. The great white shark has five very long gill slits, all located anterior to the origin of the pectoral fins. The fifth gill slit is slightly more oblique than the others, which are parallel to each other.

Like most sharks, great white sharks have eight fins, including two pectoral, two pelvic, first dorsal, second dorsal, anal and caudal fins. Except for the caudal fin, each fin has a base, anterior and posterior margin, and a tip, or apex. The posterior margin extends past the base and the lower corner of this extension is the free rear tip of the fin. The origin of a fin is the anterior-most point of the base and a fin insertion refers to the posterior-most point of the base. The caudal fin is lunate and falcate, with the upper lobe long and the lower lobe slightly shorter. The first dorsal fin is large, with a convex anterior margin, a slight concave posterior margin, and a pointed apex. How-

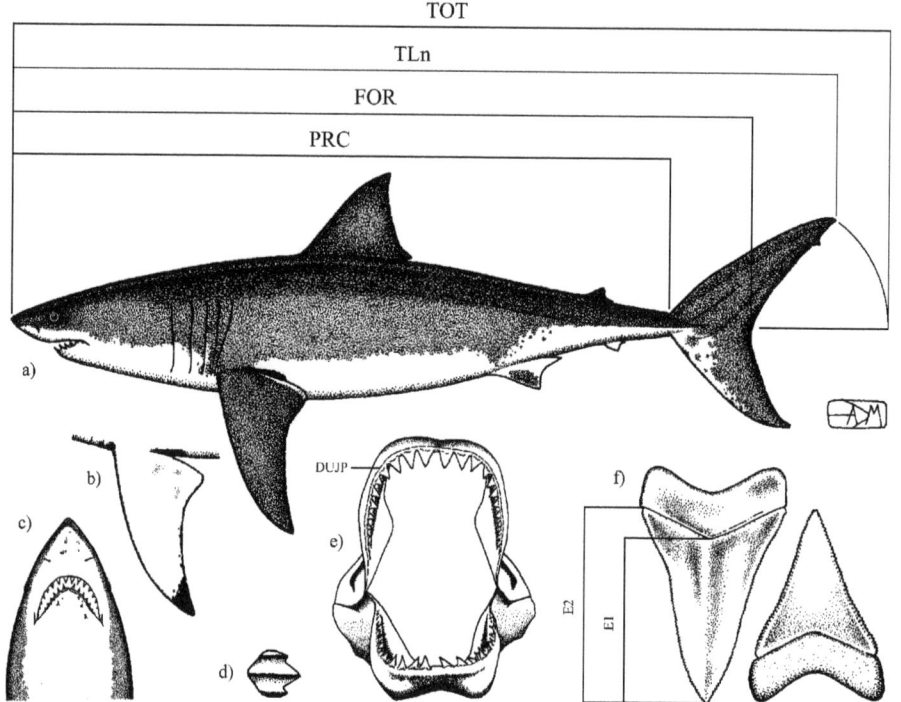

External anatomy of the great white shark and measurements used in this study: a) total length measured with the caudal fin in the natural position (TLn), total length measured with the caudal fin in the depressed position (TOT), fork length (FOR or FL), precaudal length (PRC or PL); b) ventral view of the pectoral fin; c) ventral view of the head; d) placoid scale; e) set of jaws showing the dried upper jaw perimeter (DUJP); f) upper and lower anterior teeth showing the smaller enamel height (E1) and the greater enamel height (E2) (drawing by Alessandro De Maddalena).

ever, the apex of the first dorsal fin can be rounded in newborn white sharks. Although not truly triangular, many reports of white shark sightings describe the first dorsal fin as triangular. The origin of the first dorsal fin is over the inner margin of the pectoral fins. The second dorsal fin is very small. The anal fin is approximately the same size as the second dorsal fin, with the origin of this fin slightly posterior to that of the second dorsal fin. The pelvic fins are small but larger than the anal fin. The pectoral fins are long and shaped like scythes, with their anterior margin convex, the posterior margin moderately concave and the apex pointed. The caudal peduncle, or base of the tail, is expanded laterally, forming strong dermal keels. Precaudal pits are present on the upper and lower sides of the caudal peduncle, close to the origin of the caudal fin (Bigelow and Schroeder, 1948; Compagno, 1984; De Maddalena, 2002; Last and Stevens, 1994).

The dorsal coloration ranges from deep blue to lead grey to brownish gray to almost black along the back and is a little lighter along the sides. The coloration abruptly changes to snow white on the undersides, with no color pattern. An irregular bound-

1— The Great White Shark

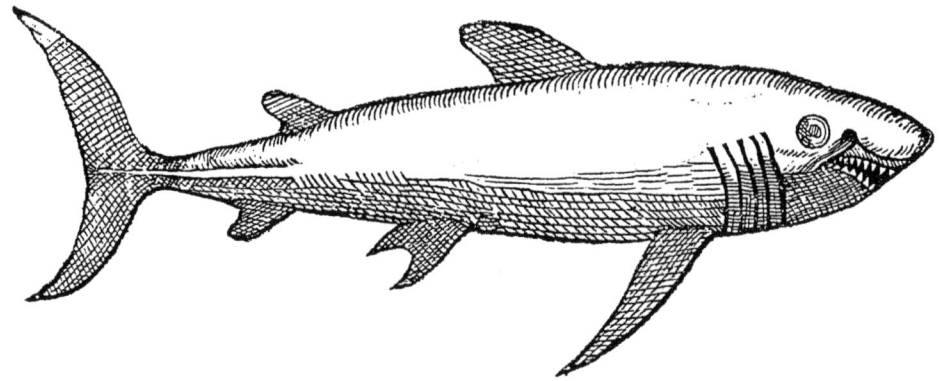

Ancient carving portraying a great white shark, from *Libri de Piscibus Marinis, in quibus verae Piscium effigies expressae sunt* by Guillaume Rondelet (illustration from Rondelet, 1554).

ary separates the dorsal dark coloration from the ventral white coloration. A black or grey area is usually present at the pectoral fin insertion, also known as the pectoral fin axil. A very distinct black mark with an irregular margin is present on the ventral surface of the pectoral fins, at the apex. An irregular whitish area is present on the anterior margin of the caudal fin lower lobe. The dark dorsal coloration partially extends to the pelvic region by forming irregular patches. Newborn white sharks have a coloration that is very similar to that of the adults. The dark coloration of the back and sides darkens to a sooty gray soon after death.

The skin of the great white shark is covered with dermal denticles that are microscopic in size, with the mean size being 0.25 mm. The skin is almost smooth to the touch, only slightly rough and abrasive. The denticles are densely distributed, and overlap each other along the anterior and lateral margins. The denticles have three crests and the posterior margin is smoothly "W" shaped. The pedicle, or the stalk-like supporting structure of the cusp, is rather short and stout (Bigelow and Schroeder, 1948).

The great white shark is one of the species of the Order Lamniformes that exhibit regional endothermy, enabling it to maintain a higher body temperature than that of the seawater because of a heat-retaining system. Red muscles are the most powerful during sustained swimming. Endothermic sharks have large amounts of red muscle tissue situated deep in the trunk, close to the vertebral column, while other species have these red muscles located closer to the skin. The red muscle tissue is connected to the circulatory system by a complicated network of arteries and veins called the "rete mirabile." The heat that is generated in the red muscles by swimming warms the blood. In the rete mirabile, the warm blood leaving the red muscle passes through the veins that carry blood from the organs to the heart. The heat is then transferred to the parallel arteries that carry cold blood from the gills to the organs. So heat is retained in the shark's body, rather than being dissipated to the environment through the gills (Ellis and McCosker, 1991). This peculiar structure of the circulatory system allows the great white shark to maintain a body temperature between 4°C and 14°C higher than the ambient water temperature (Carey et al., 1985; Goldman et al., 1996; McCosker, 1987). This higher body temperature enables the great white shark to operate at a higher metabolic rate, making this powerful predator capable of fast acceleration and high speed. This massive shark is able to leap high above the sea surface (De Maddalena, 2007, 2008).

The great white shark dental formula is usually 13–13/11–11, meaning that the great white shark usually has 13 teeth in each quadrant of its upper jaw, and 11 teeth in each quadrant of its lower jaw. The great white shark formula has variability 12 to 14–12 to 14/10 to 13–10 to 13 (Cadenat and Blache, 1981). Individual teeth of a great white shark have a large single cusp, triangular in shape and with strongly serrated margins. Great white shark teeth do not have cusplets, except in very young specimens. The teeth of newborn white sharks have two minute, pointed cusplets on each side of the cusp, and can also have a less marked serration or totally lack the serration (Uchida et al., 1996). Lower teeth are slightly smaller and narrower than upper teeth. Teeth are classified based on position in the jaw. In the upper jaw, teeth are grouped as anterior, intermediate, lateral and posterior positions. In the lower jaw, teeth are grouped as anterior, lateral and posterior (Applegate and Espinosa-Arrubarrena, 1996). Anterior teeth are longer than others, both in upper and lower jaws. The subsequent teeth are successively smaller, with the smallest teeth near the corner of the mouth. The intermediate tooth, present only in the upper jaw, is smaller than the anterior tooth that is before it and the lateral tooth that follows it. The great white shark's tooth shape is also related to its age. As the shark grows, the teeth become thick and strong to accommodate larger prey.

The great white shark has a hyostylic jaw suspension. The upper jaw does not have a direct tight connection to the chondrocranium and is loosely suspended from it by ligaments. The upper cartilages of the hyoid arch are called the right and left hyomandibulars and form a bridge attaching the jaws to the chondrocranium. As a result, the upper jaw is highly mobile and protractible (Castro, 1983; Randall, 1986). Snout elevation and upper jaw protrusion project the mouth to an almost terminal position. The bite action of the great white shark comprises a sequence of jaw and snout movements. The snout lifts, followed by a dropping of the lower jaw and protrusion of the upper jaw. The lower jaw then elevates and the snout drops, completing the bite. The entire bite action lasts about 0.9 seconds. During multiple bites, the snout remains partially lifted (Tricas and McCosker, 1984). Connected to the chondrocranium is the vertebral column that extends all the way to the tail. The total number of vertebrae of the great white shark is 172–187 (Last and Stevens, 1994). The maximum age based on vertebral band pair counts is 53 years (Sabine Wintner, personal communication).

Prey that is ingested by the white shark passes down the esophagus to the stomach. White sharks can store food in the upstream cardiac stomach after a large feed and digest it later in the downstream pyloric stomach. The partially digested food passes from the stomach to the intestinal valve, which is the ring valve that has 47 to 55 plates very close to each other (Compagno, 2001). What remains is passed downstream

1— The Great White Shark

and out of the shark through the rectum. Large undigestable parts that cannot pass to the intestine are regurgitated, as the shark can evert its stomach and expel the contents.

The white shark has acute reception capability that allows it to perceive the environment and effectively capture prey. Besides having a keen sense of smell, the great white shark has acute color vision (Gruber and Cohen, 1978). The lateral line and ears allow the shark to detect vibrations hundreds of meters away. When the shark is very close to prey, it uses its electroreception capabilities to sense it. The electroreceptors are small pores located primarily on the snout called ampullae of Lorenzini.

The Mediterranean Sea

The name Mediterranean derives from the Latin word mediterraneus, meaning "in the middle of earth" or "between lands" (medius means "middle, between" and terra means "land, earth"). This is due to the intermediary position of the sea between the continents of Africa and Europe. At the west end, the Mediterranean Sea is connected to the Atlantic Ocean and almost completely enclosed by land — on the north by Europe and Asia Minor, on the south by North Africa, and on the east by the Levant. Although the Mediterranean is technically a part of the Atlantic Ocean, it is usually identified as a completely separate body of water. The Mediterranean Sea is connected to the Atlantic

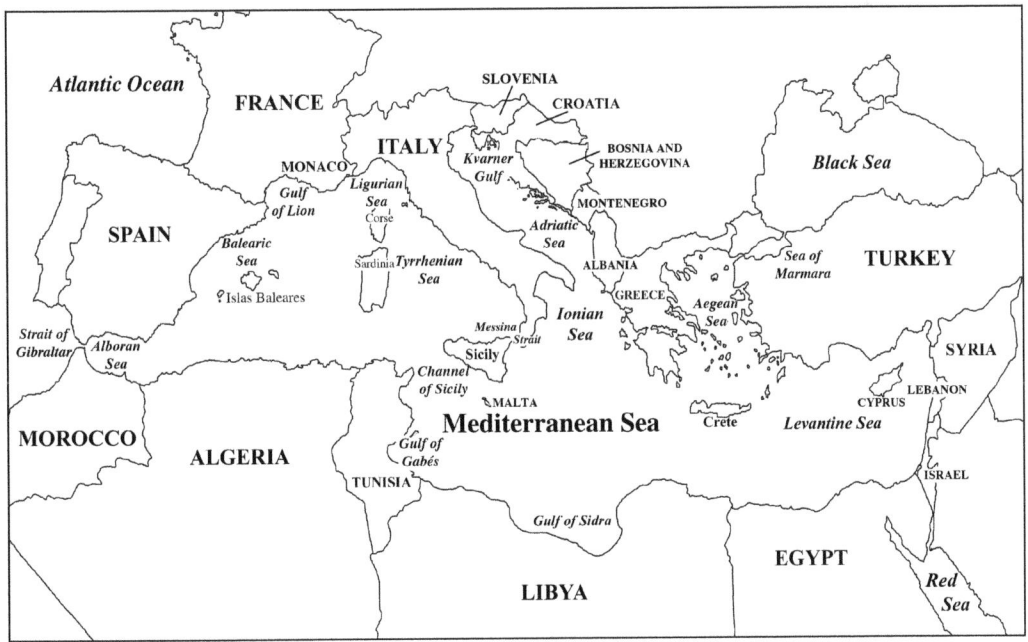

The Mediterranean Sea.

Ocean by the Strait of Gibraltar on the west and to the Sea of Marmara and the Black Sea, by the Dardanelles and the Bosporus respectively, on the east. The Sea of Marmara is often considered a part of the Mediterranean Sea, whereas the Black Sea is generally not. The man-made Suez Canal in the southeast connects the Mediterranean Sea to the Red Sea.

The Mediterranean covers an approximate area of 2.5 million km². Its connection to the Atlantic, the Strait of Gibraltar, is 14 km wide. The Mediterranean Sea has an average depth of 1500 m and the deepest recorded point is 5267 m in the Calypso Deep of the Ionian Sea. The coastline extends for 46000 km. A shallow submarine ridge, the Strait of Sicily, between the island of Sicily and the coast of Tunisia divides the sea in two main subregions, the Western Mediterranean (covering an area of about 0.85 million km²) and the Eastern Mediterranean (covering an area of about 1.65 million km²).

The Mediterranean Sea is subdivided into a number of smaller seas. The Alboran Sea is between Spain and Morocco, the Balearic Sea is between continental Spain and the Balearic Islands and the Ligurian Sea is located between Corsica and Italy. The Tyrrhenian Sea is enclosed by Sardinia, continental Italy and Sicily, and the Ionian Sea is located between Italy, Albania and Greece. The Adriatic Sea is bordered by Italy, Slovenia, Croatia, Bosnia and Herzegovina, Montenegro and Albania, and the Aegean Sea is located between Greece and Turkey. Twenty-one nations have a coastline on the Mediterranean Sea, including Spain, France, Monaco, Italy, Slovenia, Croatia, Bosnia and Herzergovina, Montenegro, Albania, Greece, Turkey, Syria, Cyprus, Lebanon, Israel, Egypt, Libya, Malta, Tunisia, Algeria and Morocco.

The climate is a typical Mediterranean climate, with hot, dry summers and mild, rainy winters. Being nearly landlocked affects the conditions in the Mediterranean Sea. For instance, tides are very limited as a result of the narrow connection with the Atlantic Ocean. Evaporation greatly exceeds precipitation and river runoff in the Mediterranean, a fact that is central to the water circulation pattern within the basin. Evaporation is especially high in its eastern half, causing the water level to decrease and salinity to increase eastward. Relatively cool, low-salinity water from the Atlantic flows eastward across the basin. As it warms and becomes saltier, it then sinks in the region of the Levant and circulates westward at depth, to spill over the Strait of Gibraltar. Thus, seawater flow is eastward in the Strait's surface waters, and westward at depth. Once in the Atlantic, this chemically distinct Mediterranean Intermediate Water can persist thousands of kilometers away from its source.

The Italian Great White Shark Data Bank

In order to collect and analyze available information about the great white sharks inhabiting the Mediterranean Sea, author De Maddalena started research in 1996 on the records of this large predator found in these waters. This data bank was initially developed as his thesis for the completion of his studies in natural sciences at the University of Milan (De Maddalena, 1997). However, the large amount of data obtained from this research, along with the interest in this project shown by many shark researchers, the general public and the media, convinced him to continue his work on this project, then named the Italian Great White Shark Data Bank. Today, this data bank includes information from 596 records of great white shark encounters from the entire Mediterranean Sea, representing the most complete and comprehensive study ever performed on the great white sharks in that area (De Maddalena, 2009). The Great White Shark Data Bank is an ongoing work. In fact, there were 43 cases added since the time this book was started. The data collected since the creation of the Italian Great White Shark Data Bank includes a substantial amount of information on size, distribution, habitat, behavior, reproduction, diet, fishery and attacks on humans that have been presented in a number of scientific articles and books (Celona et al., 2001; De Maddalena, 1998, 2000a, 2000b, 2002, 2006a, 2006b, 2009; De Maddalena et al., 2001, 2003; De Maddalena and Révelart, 2008; De Maddalena and Zuffa, 2008; Galaz and De Maddalena, 2004).

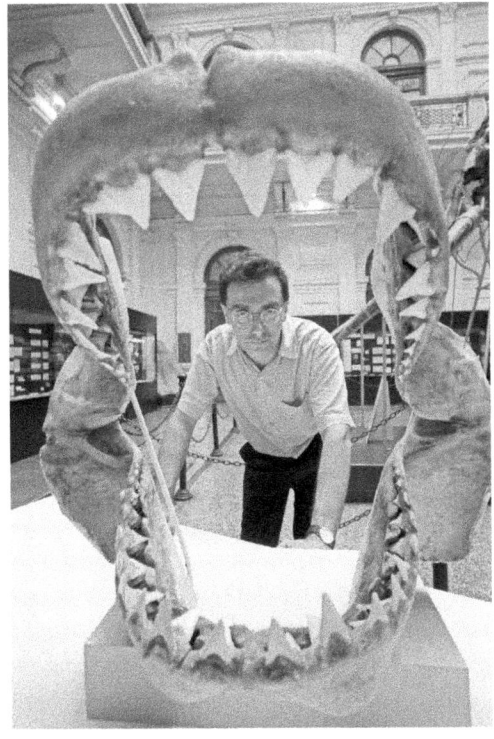

Alessandro De Maddalena examines the set of jaws from a white shark caught in the Tyrrhenian Sea on an unknown date; preserved at the Museo Civico di Storia Naturale "Giacomo Doria" of Genova, Italy (C.E. 31916) (photograph by Stéphane Granzotto, courtesy Museo Civico di Storia Naturale "Giacomo Doria," Genova).

Before the preparation of this book, the main body of information from the study was presented in four books written by De Maddalena. *Lo squalo bianco nei mari d'Italia* (Ireco, 2002) was written in Italian and focused on the presence of the great white shark along the Italian coasts. *Great White Sharks Preserved in European Museums* (Cambridge Scholars, 2007) was written in English and focused on the specimens preserved in European museums (most specimens with known capture locations come from the Mediterranean). *Le grand requin blanc sur les côtes françaises* (Turtle Prod Éditions/Média Plongée, 2008, coauthored with Anne-Lyse Révelart) was written in French and focused on the presence of the great white sharks along the French coast. *Lo Squalo Bianco nel*

Mare Mediterraneo (Rivista Marittima, 2010) was written in Italian and was a first analysis of the presence of the great white shark in the entire Mediterranean area.

The information collected in the data bank has been presented in scientific publications for specialists, but there has been a concerted effort to present the results of it to a much larger audience through a wide range of popular magazines published in Europe. This has been done in order to give accurate and current information on a species whose true nature is still largely unknown to the general public. This book represents the final stage of the study to date. All available information on records of the great white shark from the Mediterranean, including sightings from boats, underwater encounters with divers, captures by fishermen, predatory events, scavenging, attacks on humans, and museum items, is finally presented in this book. This work is ongoing and the reports received in the future will be added to the data bank.

Methods

The search for data on great white sharks from the Mediterranean Sea was facilitated by bibliographical research, location and study of materials preserved in natural history museums, visits to fish markets, collaborations with other researchers, coast guards, and private citizens, including shark enthusiasts, divers, commercial fishermen, sport fishermen, navy personnel, surfers, etc. The location and the study of white sharks preserved in natural history museums have been fundamental to this research program. Many European museums own at least some great white shark items in their collections, and some of these institutions hold remarkable specimens (a detailed report on 109 great white shark specimens preserved in 49 European institutions has been presented in De Maddalena, 2006b, 2007). The work of some researchers, and particularly the members of the Mediterranean Shark Research Group, has been particularly important in gathering new information on the presence of this species in the Mediterranean and bringing to light many old records. Among the colleagues who have contributed most actively to this study are Marco Zuffa, Tiziano Storai, Antonio Celona, Joan Barrull, Isabel Mate, Christian Capapé, Hakan Kabasakal, and Lovrenc Lipej. For a complete list of the persons who have contributed to this study from its beginning to the present, see the acknowledgments section of this book.

For present day records, the data was collected with a standard form that is shown in Appendix I. Whenever possible, the following data were collected for every case: date, location of the record, depth of the sea, depth of capture, total length (TLn or TOT) in centimeters (cm), weight in kilograms (kg), sex, stomach contents of the specimen, type of record (encounter, capture, predatory event or scavenging, attack on human or boat). Other information collected includes distance from the coast, prevailing weather and any additional details provided by the source. If available, photographs are requested. The person filing the form is the source and may refer to other sources or witnesses. This completed form is referred to as a personal communication (pers. comm.) received by the first author. However, there are other forms of personal communication, such as letters, phone conversations, or in-person interviews.

Some of the records are obtained from sources that are old and those reporting the event are no longer alive. The reports are recorded in media such as books, scientific journals, newspapers and magazines. The Global Shark Attack File (GSAF), a compilation of shark attacks worldwide based at the Shark Research Institute located in Princeton, New Jersey, is also a source of old records. Sometimes there are paintings or drawings depicting a white shark. Other records originate from public records, like payment of a bounty on white sharks. Many of these old reports are incomplete, providing only a few facts. Not all records are actual encounters with white sharks. Some records indicate the presence of a white shark. For example, a cetacean carcass with bite marks characteristic of a white shark constitutes a record, even though no white shark was actually seen. Some of the records are obtained from white sharks preserved in museums. For each white shark item preserved in a museum or institute, the following additional data was collected, if possible: institution name, institution address, storage or exhibit location, including catalogue number (cat. no.), type of material (whole specimen, set of jaws, teeth, vertebrae or other anatomical parts), preservation method (liquid, dry, taxidermy, model reproduced from a mold of a real specimen), date of acquisition, collector, and any available photographs. Morphometric measurements of the shark, teeth and jaws were taken of numerous specimens using the methods of Mollet et al. (1996). In some cases, records of white sharks are obtained from parts of sharks preserved in private collections Most of these parts are individual teeth, but in some cases they are complete jaw sets. In fact, one restaurant has two jaw sets prominently displayed for viewing by the customers.

All measurements in this book are reported in centimeters. If the original source of the data reported the shark length in feet or in another unit of measurement, the units were converted into centimeters. The weight is expressed in kilograms. If the original source of the data reported the shark weight in pounds or in another unit of weight, the units were converted into kilograms. Various measurements of size have been used in the past for the great white shark. The length has been reported both as straight line length and as measured over the curvature of the body. However, the accepted scientific measurement is straight line length. The total length (TL) is the maximum length of the shark measured from the tip of the snout to the extremity of the upper lobe of the caudal fin. The total length can be measured with the caudal fin in the natural position (TLn) or with the caudal fin in the depressed position (TOT). The fork length (FOR or FL) is the distance from the tip of the snout to the fork between the lobes of the caudal fin. The precaudal length (PRC or PL) is the distance from the tip of the snout to the origin of the caudal fin. Most of the reported lengths in the data bank are total length. In some cases the girth was reported. Girth is the circumferential measurement around the shark at its widest part, typically measured forward of the first dorsal fin. Measurements on the jaws and teeth include enamel height of the largest upper tooth (UAE1 and UAE2) and dried upper jaw perimeter (DUJP), using the methods of Mollet et al. (1996).

Methods for obtaining the length of great white sharks from commonly preserved skeletal parts (teeth, jaws, vertebrae) have been investigated by various authors (De Maddalena, 2000b, 2002; Gottfried et al., 1996; Mollet et al., 1996; Randall, 1973, 1987; Zuffa et al., 2002). Mollet et al. (1996) and De Maddalena (2007) found that the size of the largest teeth is a reliable index for estimating the size of a very young shark, but it cannot be used reliably to estimate the size of medium or large individuals. There is also a relationship between the size and the perimeter of the upper jaw, but it has been considered inaccurate for length predictions (De Maddalena, 2000b, 2002, 2006b; 2007, 2009; Mollet et al., 1996). Consequently, these methods have not been used in this work. The size of a great white shark can be estimated from the diameter of the vertebral centrum. However, there is insufficient evidence supporting this method as reliable. Considering the failure of the attempts to estimate the size of a white shark based on the size of the teeth and jaws, it is doubtful that accurate correlations for length can be obtained from the utilization of other skeletal parts. Given the rarity of cases in which the largest vertebral centra of a great white shark of unknown size was preserved, the correlation between white shark size and vertebral centra size is not useful. In conclusion, there is currently no reliable method to estimate the size of a white shark from parts that are commonly preserved.

Photographic documentation has sometimes been used to estimate the size of large white sharks, and then to validate or refute the size reported by the original source or eyewitnesses. The literature on great white sharks is rich with gross estimates obtained from photos, which are unsuitable for scientific evaluation (De Maddalena, 2009). As it has been pointed out in De Maddalena et al. (2001), the photos must be used with great caution to estimate great white shark size, since the probability of obtaining inaccurate results is high. It is absolutely necessary to use only pictures where the whole specimen or a great part of it is not deformed by imperfections in the camera lens or by the perspective of the shot. Suitable photographs are rare. It is also essential that a large object of known size is located in the same plane of the shark in the photograph. Photos that do not meet this criteria have to be considered of no value for producing a valid estimate of the specimen size.

The data bank has several sharks with a reported length between 500 cm and 600 cm. There are many reported in excess of 600 cm. A 589 cm TOT female great white shark captured off Maguelone and landed in Sète, France, on October 13, 1956, is the largest specimen for which complete morphometric measurements are available (De Maddalena et al., 2003). This morphometric data has been used in this study to produce reliable estimates of large white sharks of equivalent size, approximately 600 cm in length, when suitable photographic documentation existed (De Maddalena et al.,

2001, 2003). The names of the locations used in this work are in their original languages. Regarding the country boundaries of the white shark records, in this book, the present day boundaries were referenced. For example, some sightings, captures and attacks were recorded along the present day Croatian coast, which was an Italian territory at the time. The present day boundaries were considered and these cases were recorded from Croatia.

2

Records of Great White Sharks from the Mediterranean Sea

Spain

Great white shark captures and sightings have been reported from both continental Spain and off the Islas Baleares.

Continental Spain

The earliest record in the data bank for continental Spain was in 1878, when a great white shark was caught in a net between Islas Columbretes and Castellón. It was reported to measure about 500 cm and to weigh 2300 kg (Barrull and Mate, 2001; Lozano-Rey, 1928; Perez-Arcas, 1878). Around 1920, a 150 cm specimen was caught off Melilla and displayed at the fish market in Melilla (Fergusson, 1996). In 1927 (the date may be approximate) a specimen was caught off Premiá de Mar (Francesc Xavier Viñals Moncusi, pers. comm.). The next record was over 40 years later in 1962. A specimen was caught off Castellón and was reported measuring 500 cm and weighing 1500 kg. Its stomach contained a tuna and a dolphin (Asensi, 1977; Barrull, 1993–1994; Francesc Xavier Viñals Moncusi, pers. comm.).

On March 18, 1983, or 1986 (or in June, depending on the source), an estimated 350 cm great white shark was reported attacking windsurfer L. Pérez-Díaz at 10:55 a.m., about 300 m off Tarifa. The windsurfer survived but lost a foot. It has been suggested by some that the species responsible could have been a shortfin mako *Isurus oxyrinchus*, while others considered this case doubtful (Joan Barrull, pers. comm.; Francesc Xavier Viñals Moncusi, pers. comm.; GSAF). On November 17, 1992, a 475 cm male was seen swimming slowly for at least six hours just below the surface of shallow coastal waters off Tossa de Mar. After midnight the shark was stranded and moribund on the Playa de la Mar Menuda, where it died one hour later. It was reported weighing 1000 kg, even though there were no contents in the stomach. The jaws are preserved at the Centro de Recuperación de Animales Marinos Fundación CRAM in Premiá de Mar. Photographic and film documentation of this specimen exists. Barrull and Mate (2001) reported this record on the basis of a personal communication they received from Ferran Alegre (Barrull, 1993, 1993–1994; Barrull et al., 1999; Barrull and Mate, 2001; Francesc Xavier Viñals Moncusi, pers. comm.). Francesc Xavier Viñals Moncusi (pers. comm.) reported a different date, 1993 instead of 1992.

In addition to the cases mentioned above, there are some cases lacking an exact

2—Records of Great White Sharks from the Mediterranean Sea

This estimated 550 cm white shark was caught on August 10, 1946, near Tabarca, Spain (photograph by Sánchez, AMA, courtesy Alicante Vivo.org).

date and are recorded as having occurred before a reference date from the existing documentation. For example, if the teeth were donated to a museum on a certain date, that date becomes the reference and the specimen is said to be caught prior to that date. Before 1878, a capture may have occurred off Puerto de Mazarrón, but this case is considered doubtful (Francesc Xavier Viñals Moncusi, pers. comm.; Perez-Arcas, 1878). Before 1878, a 500 cm specimen was caught off Valencia (Perez-Arcas, 1878). This case is considered doubtful, since it may be the same capture that occurred in 1878 between Islas Columbretes and Castellón. Before December 16, 1912, a 471 cm specimen was caught in a tuna trap off Vilassar de Mar. The reported weight was 1000 kg. Two teeth from that specimen are preserved at the Museu de Zoologia in Barcelona, Spain (cat. no. MZB-82-5316 and MZB-82-5317), which were donated to the museum by Joan Prim on December 16, 1912 (Barrull, 1993–1994; Barrull et al., 1999; Barrull and Mate, 2000). Before 1926, a specimen was reported caught off Tarragona. According to Francesc Xavier Viñals Moncusi (pers. comm.), its teeth should be preserved at the Museu de Zoologia in Barcelona, Spain, but according to Barrull et al. (1999) this record does not exist (Barrull et al., 1999; De Buen, 1926; Fergusson, 1996; Francesc Xavier Viñals Moncusi, pers. comm.). Prior to 1928, a specimen was caught off Vinaroz. The jaws presumably are preserved at the Museo Nacional de Ciencias Naturales in Madrid, Spain (Lozano-Rey, 1928).

On August 10, 1946, a large white shark was caught in a tuna trap near Tabarca, off Alicante (http://www.lafogueradetabarca.blogspot.com). Its stomach contained a 40 kg tuna. Photographic documentation of this shark exists. The photo does not allow an accurate estimation of the shark size, but De Maddalena grossly estimated it to be approximately 550 cm. Before 1999, a specimen was caught on a longline off Alboran (Francesc Xavier Viñals Moncusi, pers. comm.).

Islas Baleares

Prior to the 1980s, tuna were fished off Islas Baleares using traps. These traps also caught several white sharks. The earliest record in the data bank for Islas Baleares is from an unspecified date in the 1920s, where a great white shark was caught in a tuna trap located off the northeastern coast of Majorca. It was estimated to be 390 cm TL by Morey et al. (2003) on the basis of the photographic evidence that still exists. Morey et al. (2003) reported this record on the basis of a personal communication they received, by G. Blanc. In another unspecified date in the 1920s, a specimen was caught in a tuna trap located off the northeastern coast of Majorca. It was reported weighing 2000 kg. Morey et al. (2003) reported this record on the basis of personal communication from F. Riera. A third specimen was caught in a tuna trap located off the southeastern coast of Majorca on an unspecified date in the 1920s. It was reported to be 700 cm in length. The record of this large shark was reported by Morey et al. (2003) based on personal communication from L. Vadell. On September 3, 1927, a specimen was caught in a tuna trap located off the southwestern coast of Majorca. It also measured 700 cm as reported. Morey et al. (2003) based this record on personal communication with J. Morey. Still another large shark measuring 700 cm was caught in a tuna trap in the winter of 1935 off the southwestern coast of Majorca. Morey et al. (2003) reported this record on the basis of a personal communication they received, by B. Ginard.

The 1940s documented many great white sharks caught in the tuna traps. In a winter during the 1940s, a specimen was caught in a tuna trap located off the northeastern coast of Majorca. It was reported over 400 cm in length and weighed 800 kg. Morey et al. (2003) reported this record on the basis of personal communication with M. Cerdà. Another specimen was caught in a tuna trap located off the northeastern coast of Majorca in a winter during the 1940s. It also was reported to measure over 400 cm and weighed 1000 kg. Morey et al. (2003) reported this record on the basis of a personal communication with M. Cerdà. On February 1, 1942 a specimen was caught in a tuna trap located off the northeastern coast of Majorca. Like the previous two, this shark was reported at 400 cm in length and weighed 800 kg. This record was based on personal communication from J. Borràsbu as reported by Morey et al. (2003). This specimen is most likely the same specimen reported by Ramis (1988) that had a common thresher shark *Alopias vulpinus* in its stomach, even though the date and weight reported by this source are different (it is reported as having been caught in 1941 and weighing

900 kg). On February 12, 1944, a large specimen measuring 535 cm in length was caught in a tuna trap located off the northeastern coast of Majorca. It was reported weighing 1350 kg. Morey et al. (2003) reported this record on the basis of a personal communication with J. Borràs.

There were many white shark captures from Islas Baleares from the 1960s as casualties of the tuna trap fishery. In March 1962, a 350 cm specimen was caught in a tuna trap located off the northeastern coast of Majorca. It was reported to weigh 500 kg and documented with photographs, which still exist. Morey et al. (2003) reported this record on the basis of personal communication from J. Borràs. In the same month, a 300 cm specimen was caught in a tuna trap located off the northeastern coast of Majorca. It was reported to weigh 300 kg and, like the previous record, photographic documentation of this specimen also

One of the white sharks caught in 1969 off Majorca, Spain (courtesy Brisas).

exists. Morey et al. (2003) reported this record on the basis of a personal communication from J. Borràs. In the winter of 1963, a 500 cm specimen was caught in a tuna trap located off the northeastern coast of Majorca. It was reported weighing 500 kg. This record was reported by Morey et al. (2003) and is based on personal communication from A. Salas. In the same winter, on December 26, 1963, a large 615 cm female shark was caught in a tuna trap located off the northeastern coast of Majorca. It was reported to weigh 2200 kg. This huge specimen was photographed and the documentation still exists. This record is from a personal communication from J. Borràs as reported by Morey et al. (2003). In January 1964, another large female shark was caught in a tuna trap located off the northeastern coast of Majorca. It was reported measuring 535 cm and weighing 1400 kg. However, it was estimated to be 510 cm TL by Morey et al. (2003) on the basis of the photographic evidence that is still in existence. Morey et al. (2003) reported this record on the basis of a personal communication from J.

Borràs. A year later, in the winter of 1965, another large shark was caught in a tuna trap located off the northeastern coast of Majorca, that was reported measuring 700 cm and weighing 1000 kg. Morey et al. (2003) reported this record based on personal communication from S. Bisbal.

In 1966, a female shark reported to weigh 1100 kg was caught in a tuna trap located off the northeastern coast of Majorca. Another female shark was caught in 1966 in a tuna trap located off the northeastern coast of Majorca. It was slightly larger than the previous shark and reported to weigh 1200 kg. For both sharks, Morey et al. (2003) reported this record on the basis of a personal communication they received from J. Borràs. A year later, in January 1967, another large female shark was caught in a tuna trap located off the northeastern coast of Majorca. It was reported weighing 1700 kg. Based on photographic documentation, it was estimated to be 550 cm TL by Morey et al. (2003). This record was based on personal communication from O. Pinet as reported by Morey et al. (2003). Eight months later, on October 1, 1967, a male shark was caught in a tuna trap located off the northeastern coast of Majorca. It was reported measuring 515 cm and weighing 1700 kg. Photographic documentation of this specimen exists and it was estimated to be 450 cm TL by Morey et al. (2003) based on the photographic evidence. Morey et al. (2003) reported this record on the basis of a personal communication they received by J. Rullan. In the same year, a 535 cm female was caught by Francisco Pérez in a tuna trap located off the northeastern coast of Majorca. This shark was reported to weigh 1350 kg. Photographs and the documentation of this specimen exists. Morey et al. (2003) reported this record on the basis of a personal communication with the captor, F. Pérez. This is probably the same specimen reported by Ramis (1988) that had a dolphin and four large tunas in its stomach, even though the weight reported by this source is slightly higher (1500 kg) and is simply indicated as having been caught before 1976.

Four sharks were caught in the tuna traps in 1969. In February of 1969, a large female was caught in a tuna trap located off the northeastern coast of Majorca. It was reported measuring 550 cm and weighing 2000 kg. However, it was estimated to be 535 cm TL by Morey et al. (2003) on the basis of the photographic evidence that still exists. Morey et al. (2003) reported this record on the basis of a personal communication they received, by J. Domingo. A second shark was captured in a tuna trap on February 1969 off the northeastern coast of Majorca. The reports show the shark weighing 1250 kg. Based on photographic documentation of this specimen, the length was estimated to be 550 cm TL by Morey et al. (2003). This record was based on personal communication that Morey et al. (2003) received from A. Vera. In February 1969, a third specimen was caught in a tuna trap located off the northeastern coast of Majorca. This shark was reported to weigh 1000 kg and Morey et al. (2003) reported this record based on personal communication they received from A. Vera. Two months later, in March of 1969, a large female shark was caught in a tuna trap located off the southwestern coast of Majorca. Per the report, it measured 800 cm and weighed 2500 kg.

However, photographic evidence of this specimen provided an estimate of 620 cm TL by Morey et al. (2003), who reported this record on the basis of a personal communication they received by G. Ferragut.

Three sharks were reported in the 1970s. In January 1970, a specimen was caught in a tuna trap located off the northeastern coast of Majorca. It was reported to weigh 1350 kg. Morey et al. (2003) reported this record based on personal communication with M. Alberti. In 1972, another large shark was caught in a tuna trap located off the northeastern coast of Majorca and weighed 2500 kg per the report by Morey et al. (2003) based on personal communication they received from M. Alberti. On February 5, 1976, a large female shark was caught in a tuna trap located off the northeastern coast of Majorca. This large shark was measured at 615 cm and weighed 2500 kg. The length was confirmed at 610 cm TL based on photographic documentation by Morey et al. (2003). Morey et al. (2003) reported this record on the basis of personal communication from F. Pérez. This is probably the same specimen reported by Ramis (1988) and Marco Zuffa (pers. comm.) that had a large manta or ray in its stomach, even though the date reported by this source is slightly different (February 10, 1976).

After 15 years without a record, two specimens were reported caught in July 1992 by a fisherman off Andraitx, Majorca. They were reported measuring 500 cm each. Fergusson (1996) reported this record on the basis of a personal communication he received from J. Piza. But according to Barrull and Mate (2001), this case has to be considered highly doubtful since J. Piza did not report this capture and it was impossible to find a single fisherman in Andraitx that would confirm this case.

After 1992, there are no recorded captures in the data bank of white sharks from Islas Baleares. The reports are evidence of white shark presence. On July 5, 1997, the carcass of an 8 m sperm whale, *Physeter macrocephalus*, was found stranded at the Islas Baleares bearing white shark bites (Morey et al., 2003). In 1998, three more carcasses were found stranded that showed evidence of white shark presence. The first was on March 21, 1998, where the carcass of a 2.2 m striped dolphin, *Stenella coeruleoalba*, was found stranded at the Islas Baleares bearing white shark bites in the urogenital region (Morey et al., 2003). Three months later on June 13, 1998, the carcass of a loggerhead sea turtle *Caretta caretta*, was found stranded at the Islas Baleares, also bearing white shark bites on the pelvic flippers, carapace and plastron (Morey et al., 2003). The third was on September 8, 1998, where a carcass of a loggerhead sea turtle *Caretta caretta*, was found stranded at the Islas Baleares bearing white shark bites on the plastron (Morey et al., 2003). Approximately two years later on July 23, 2000, the carcass of a 3.3 m bottlenose dolphin *Tursiops truncatus*, was found stranded at the Islas Baleares bearing white shark bites in the urogenital region (Morey et al., 2003). Seventeen days later on August 9, 2000, another carcass of a 3.5 m bottlenose dolphin *Tursiops truncatus*, was found stranded at the Islas Baleares, bearing white shark bites in the urogenital region (Morey et al., 2003). In the same month, an estimated 400 cm white shark was reported off Cabrera. The shark was seen at nightfall by fishermen. A military patrol was dispatched and reached the spot, but could not confirm the sighting (Barrul and Mate, 2001). In the following year two more dolphin carcasses were found. On March 3, 2001, the carcass of a Risso's dolphin *Grampus griseus*, was found stranded at the Islas Baleares bearing white shark bites and on April 19, 2001, the carcass of a 1.9 m striped dolphin *Stenella coeruleoalba*, was found stranded at the Islas Baleares bearing white shark bites in the urogenital region (Morey et al., 2003).

In addition to all the cases described above, there is a case without an exact date and was recorded as having occurred prior to the first mention in the existing documentation. Before 1868, a specimen was recorded off Minorca, but it is unclear if the specimen was captured (Barcelo and Combis, 1868; Fergusson, 1996).

France

Great white shark captures and sightings have been reported from both continental France and off Corsica.

Continental France

The earliest records reporting the presence of the great white shark along the French coast of the Mediterranean Sea date from the Middle Ages. Gianturco (1978) wrote that in the Middle Ages chronicle of Aix-en-Provence, a man wearing armor was found in the stomach of a great shark. Surely Smith (1833) referred to the same shark when he wrote that in the records of Aix there was an account of a 670 cm shark taken by fishermen in which the stomach contained, among other undigested remains, the headless body of a man encased in complete armor. Unfortunately, both Smith (1833) and Gianturco (1978) did not specify the source of this record nor its date (the Middle Ages range from 476 to 1453) nor the exact location where the specimen was caught. Rondelet (1554), probably referring to the same case when writing about the presence of the great white shark along the French coast of the Mediterranean, stated that white sharks with a whole man clad in armor in their stomachs were caught in Nice and in Marseille. The case of a man found in the white shark caught in Marseille was also reported by Gessner (1560).

In the 18th century, some cases were reported, but only limited details were provided. In the stomach of a white shark caught between Cassis and La Ciotat, two tuna and an entire dressed man were found. Off Antibes, a bathing seaman was attacked and had a leg severed cleanly by the jaws of a white shark while his companions pulled him out of the water. In Nice, it was reported that a child was devoured by a white shark. Off Cannes, near Île Sainte-Marguerite, a large specimen was caught that had an entire horse in its stomach (Cazeils, 1998).

Besides reports of great white sharks, one was featured in artwork. In Lacépède (1839), there is a color engraving by the Swiss illustrator Charles-Joseph Traviès de

Villers, entitled *Le requin*, portraying a freshly caught huge female white shark on a beach, possibly on the French coast of the Mediterranean. Even though the illustration has some imperfections, many details appear to be extremely accurate, suggesting that Traviès has without any doubt portrayed a genuine white shark.

Around 1860, the presence of the white shark was reported in the waters of Sète by Doumet (1860), but the report did not specify any particular capture or sighting. In May 1861, a white shark was caught in Sète, and a row of teeth of this specimen is preserved at the Muséum National d'Histoire Naturelle in Paris (cat. no. MNHN ab-0195) (De Maddalena, 2006). In August 1875, an approximately 400 cm white shark was caught off Sète. It measured 223 cm in girth and weighed approximately 600 kg (Moreau, 1881). A year later, in 1876, a 242 cm specimen was caught off Sète. It measured 150 cm in girth. The source reported many other morphometric measurements (Moreau, 1881). In 1888, a 250.5 cm male white shark was caught at the Grau-du-Roi. This specimen is preserved as a taxidermied skin-mount at the Muséum d'Histoire Naturelle de Nîmes (cat. no. BAC 132) (De Maddalena, 2006). A year later, in 1889, a specimen was caught off La Seyne-sur-Mer. Four of its teeth are preserved at the Muséum National d'Histoire Naturelle in Paris (cat. no. MNHN ab-0185) (De Maddalena, 2006, 2007). It is possible that this case is the same one that follows. In October 1889, a 400 cm specimen was caught in the nets of fishermen from Le Brusc. Its stomach contained a porpoise and the legs and pelvis of a man (Dujardin, 1890; Moreau, 1892). Carus (1893) reported the presence of great white sharks in the waters of Toulon and the capture of some specimens off Nice, but without any additional detail. Shortly before 1898, a 550 cm great white shark was caught in a net in the Var region. This large shark was reported having a girth of 400 cm and weighing approximately 2000 kg. Its stomach contained a porpoise (*Phocoena phocoena*), three tuna and other fishes. According to Bonomi (1898), the specimen was probably shipped to the Muséum National d'Histoire Naturelle in Paris.

The twentieth century provided many of the reports from France in the data bank. Two specimens were caught on an unknown date before 1909 in the region of Camargue or of Martigues. Their sets of jaws are preserved at the Musée Océanographique de Monaco (cat. no. MOM-P0I-4254 e MOM-P0I-4253, previous catalogue number: 911). The museum bought them on October 13, 1909, from the owner of a bar on cours Saleya in Nice (Bruni and Würtz, 2002; De Maddalena, 2006; Roule, 1912). On October 11, 1910, a specimen was caught off Grau-du-Roi. Three of its teeth are preserved at the Muséum d'Histoire Naturelle de Nîmes (cat. no. 6) (De Maddalena, 2006). Approximately two years later, in 1912, a large specimen was caught by a fisherman from Le Brusc off Six-Four-les-Plages (Gérard Altman, pers. comm.; De Maddalena and Zuffa, 2008). Thirteen years later, on October 15, 1925, a female was caught by A. Rouard at the Estaque, a quarter of Marseille. It was reported measuring 600 cm in length and weighing 1500 kg. Photographic documentation of this specimen exists (De Maddalena and Zuffa, 2008; Philippe Summonti, pers. comm.). The photo does allow an accurate estimation of the shark size, so author De Maddalena estimated it to be 667–687 cm TOT. Ten years later, in the summer of 1935, a 500 cm specimen was caught off pointe de l'Espiguette. It was reported weighing 1600 kg gutted and its stomach contained three whole undigested tunas (Perrier, 1938). Another shark was caught in the summer of 1935 and others were sighted in the waters off the Languedoc region (Perrier, 1938). A very large specimen was caught around 1940 off Grau-du-Roi that was reported measuring 800 cm in length. Granier (1964) wrote that it was roughly stuffed and exhibited by a stall keeper throughout the whole Provence region. Other specimens were caught in 1943 and 1946, between Palavas and Grau-du-Roi (Granier, 1964). Jacky Granier conserved a tooth from the specimen caught in 1946 (Marco Zuffa, pers. comm.). During the 1950s, a white shark was reported to have bit off the feet of two fishermen off Toulon (Touret, 1992).

On October 13, 1956, a large female shark was caught off Maguelone and landed in Sète. The local newspaper, *Midi libre*, published the details of the capture (Anonymous, 1956). Some photos, showing the shark surrounded by a crowd, were published in this article. The shark was caught aboard the trawler *Rosina-Raphaël*, which was owned by M. Antoine Ferrignon and crewed by M.M. Priore Charles, Priore René, Borozzi Marcel, and Henric Gaëtan. The *Rosina-Raphaël* was fishing about three miles off Maguelone when one of the sailors noticed that a part of the 120 m long net was at the bottom. The crew hauled up the sunken part of the net, which had entangled a huge fish. After a brief struggle, the large shark died. The master towed it to Sète, arriving around nine o'clock in the morning. The shark was hauled up on the quay with considerable difficulty, requiring the help of a mast and a tractor. The monster shark attracted a large crowd as it was displayed in front of the store of M. Azaïs, a fish wholesaler. The shark was reported measuring 589 cm and having a maximum girth of 400 cm. Its weight was estimated to be about two tons and its liver alone weighed 360 kg. The remains of two dolphins measuring about 180 cm were found in the stomach. The huge fish was examined by M. Euzet, a lecturer at the Marine Biological Station of Sète and by M. Baer, a professor at the University of Neuchatel, Switzerland. Baer immediately contacted the Musée d'Histoire Naturelle in Lausanne, which decided to purchase the shark and have it mounted. The shark was transported to Lausanne in a freezer wagon, but it was not frozen properly and arrived partially spoiled.

The shark was cut in pieces at the slaughterhouse in Malley and a cast was taken of each piece. Unfortunately, the skin was too damaged and could not be used. Of the original shark carcass, only the fins and teeth were used in the final model. A metal structure was constructed to support the cast portions of the shark. The model was then prepared by taxidermist Eugène Küttel. Even though the model prepared by Küttel showed some deformities on the snout, lower jaw and ventral surface of the trunk, the taxidermist accurately reproduced the original specimen in most respects. The finished model showed some deformities, especially in the head, which can easily be attributed to the natural deterioration to which the shark was subjected after its death and the less

This engraving by the Swiss illustrator Charles-Joseph Traviès de Villers, entitled *Le requin*, portrays a huge female white shark on a beach, possibly on the French coast of the Mediterranean (illustration from Lacépède, 1839).

than ideal conditions of the transport. The ventral portion of the shark was also deformed, probably because it was gutted prior to arrival in Lausanne (De Maddalena et al., 2003). In reality, large sharks take an unnatural shape out of the water. Modern day taxidermists correct these deformities by applying a filler to restore the true shape (De Maddalena and Heim, 2009). This large shark is on exhibit at the Musée Cantonal de Zoologie in Lausanne (without a catalogue number). Three of the original teeth have been replaced with replica teeth because they were stolen. In the summer of 2005, the model was cleaned and restored by painter Olivier Besse and taxidermist André Keiser (De Maddalena and Révelart, 2008).

Complete morphometric measurements of this shark are available (De Maddalena et al., 2003). The 583 cm total length (TOT) of the model confirmed the 589 cm total length that was reported by "Anonymous" (1956) for the freshly captured specimen (1956). Other authors mentioned this specimen, reporting erroneous sizes: Beaumont (1957) wrote that it measured around 5 m and weighed at least 1.5 tons; Quignard et al. (1962) reported that it measured 490 cm; and Fergusson (1996) wrote that it was 4.5 m long and also reported a wrong date of 1976. The specimen is the largest cast in the world that was reconstructed directly from a whole specimen. The shark was accurately measured at the time of capture and is still measurable today from the cast, making it the largest specimen ever reported with a size that cannot be disputed (De Maddalena et al., 2003).

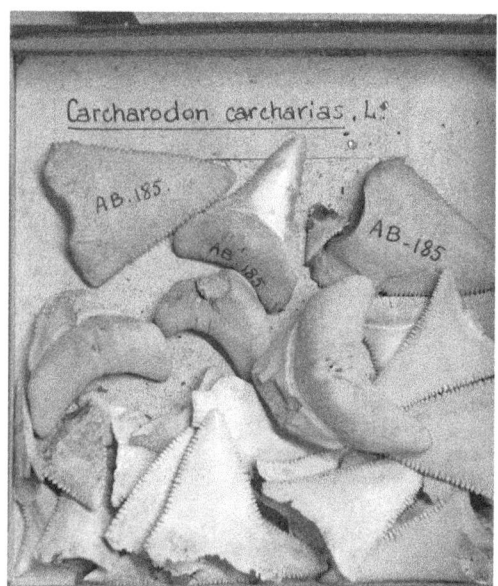 

Left: These teeth from a white shark caught in 1889 off La Seyne-sur-Mer, France, are preserved at the Muséum National d'Historie Naturelle in Paris, France (numéro d'inventaire: MNHN ab-0185) (photograph by Bernard Séret, courtesy Muséum National d'Historie Naturelle de Paris). *Right:* Allessandro De Maddalena's wife, biologist Alessandra Baldi, showing the set of jaws from an estimated 550 cm white shark caught before 1909 in Camargue, France, and preserved at the Oceanographic Museum of Monaco (cat. no. MOM-P0I-4254) (photograph by Alessandro De Maddalena, courtesy Musée Océanographique de Monaco, Monaco–Ville).

A second shark was caught in 1956 between Palavas and Grau-du-Roi (Granier, 1964). That same year there was even a third shark caught, 200 m offshore in a location called *La Queue*, between les calanques of Niolon and Figuerolles. This shark was caught in a drift net that was intended to catch tuna. Among the three fishermen from the Estaque, a quarter of Marseille, who took part in this capture were the master, Mr. Scotto (nicknamed *Le noir*) and Bernard Escobar. The shark was reported weighing 1800 kg and its stomach contained the remains of several dolphins, large tunas and a monk seal pup. Photographs of this specimen exist. The specimen was shipped to Paris for exhibit (Damonte, 1993; Perosino, 1963).

During the 1960s and 1970s only one shark report is in the data bank for France. On an unknown date prior to 1965 a white shark was reported to attack a worker who fell into the turbid and shallow waters of the harbor of Marseille (Hemingway and Devlin, 1965).

After more than a decade with no records, white shark presence was again reported in the 1980s. On December 4, 1986, two firemen of the Battalion of Marseille encountered a great white shark near Îles du Friol. On December 6, 1986, the news was published in the newspaper *Le Provençal*, which reported the case as an attack (Anonymous, 1986). In reality, it was not an attack. On the morning of December 4, 1986, the motor-

 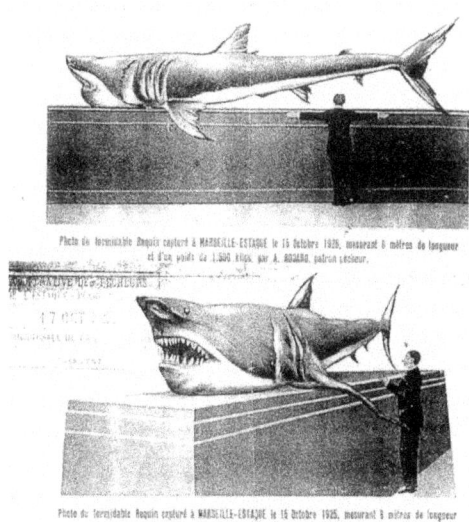

Left: This set of jaws from a great white shark caught before 1909 in Martigues, France, is preserved at the Oceanographic Museum of Monaco (cat. no. MOM-P0I-4253), Monaco (photograph by Alessandro De Maddalena, courtesy Musée Océanographique de Monaco, Monaco–Ville). *Right:* This 600 cm white shark was caught on October 15, 1925, off Estaque, Marseille, France (courtesy Philippe Summonti).

boat *Boufareu* received an order to conduct a training dive in the Sector Tiboulen Frioul. The crew consisted of master major Aimone, second master second major Alziari, petty officer third class Ferrand, and the divers, Major Ciabrini, petty officer third class Cupif and petty officer third class Faure. At 9:45, the motorboat *Boufareu* left from the harbor of Pointe Rouge en route to Tiboulen of Frioul and arrived on the spot at 10:10. Since the weather was very good, the crew decided not to drop anchor.

The training was required to take place as follows: Alain Alziari and Marc Ferrand dove first, followed by the other divers five minutes later. At 10:24 the first two divers entered the water and descended toward the bottom at a depth of 28 m. Ferrand led the way 20 m in front of Alziari. The visibility was reported as very good. As he descended, Alziari was looking at the bottom directly below him becoming closer and closer. He looked away for a moment and to his astonishment, he saw a huge shark swimming leisurely near the bottom. At first, he mistook the animal for a basking shark, but he soon realized that it was a great white. Alziari immediately searched for his friend and spotted him at the bottom some distance away. Ferrand was unaware of the shark behind him. The only way for Alziari to warn his friend was to make some noise, but that would risk attracting the attention of the shark to himself. Therefore, he did nothing. Alziari watched the shark approach Ferrand and just before any possible contact, Ferrand turned around to look for Alziari. He immediately saw the shark swimming toward him. Pushing on the bottom, Ferrand launched himself upward and the shark passed between his legs, with the predator's first dorsal fin brushing his thigh. Both divers ascended toward the surface, Alziari surfacing first.

As Ferrand ascended, the shark came back around and Ferrand saw it pass below his flippers. Ferrand looked above him and saw the surface was very close. Then he looked below him and saw the shark coming up from the bottom toward him, its mouth wide open and its eyes rolled backward. Ferrand surfaced at the same moment that Alziari was pulled aboard the boat. Ferrand hit the water, shouting and purging his regulator. The shark closed its mouth, turned and disappeared into the deeper waters and Ferrand was immediately taken on board. A few moments later, the crew spotted the shark and were amazed. It made several circles around the *Boufareu,* enabling the crew to estimate its length at 500–600 cm. Then the shark disappeared. The *Boufareu* warned other boats in the area about the presence of the shark. At 11:00, the boat motored in the direction of the harbor of La Pointe Rouge, in Marseille. Master Major Aimone reported the incident to the sergeant major of La Pointe Rouge (Alain Alziari, pers. comm.; De Maddalena and Révelart, 2008; De Maddalena and Zuffa, 2008).

This 589 cm female white shark captured off Maguelone, France, on October 13, 1956, is preserved as the largest cast in the world, which was reconstructed directly from a whole specimen and is located in the Museum of Zoology in Lausanne, Switzerland (without a catalogue number) (photograph by Michel Krafft, courtesy Musée cantonal de Zoologie, Lausanne).

In 1989, a large white shark was sighted by fishermen following a school of tuna. This specimen was estimated to measure over 400 cm and photographs were taken to document the encounter, which still exist (Fergusson, 1996; Touret, 1992). More white shark encounters were reported in the 1990s. Around 1990, a white shark was encountered by M. Arthur, a technician at the Entente Interdépartementale pour la Démoustication (EID), while he was fishing off Valras, near the mouth of the Aude river. The shark raised its head out of the water to investigate the boat and its occupant (Guy Oliver, pers. comm.).

On January 9, 1991, another large white shark was caught off Sète. The trawler *Jean Licciardi* from Sète was 45 miles offshore retrieving its net when the crew found a female white shark trapped in it. The already dead shark was hoisted on the deck of the trawler and later landed at the Quai de la Marine in Sète. A crowd gathered to see

Top: This white shark was caught in 1956 between Niolon and Figuerolles, France (photograph by Agence Intercontinentale, Paris). *Bottom:* This estimated 591 cm white shark was caught on January 9, 1991, off Sète, France (photograph by Raymond de Neuville, courtesy Midi Libre).

this large shark that was estimated to weigh around two tons. The stomach contained the remains of four dolphins measuring from 80 to 100 cm and two swordfish, *Xiphias gladius*. The six sailors removed and preserved the teeth as a trophy. Jean Licciardi, the boat master, and his crew posed with the large shark for photographer Raymond de Neuville. A fish wholesaler bought the carcass and put it up for sale at the fish market in Rungis. It was later purchased by a supermarket in Montargis (Anonymous, 1991b;

De Maddalena et al., 2001, 2003; Quignard and Raibaut, 1993; Séret, 1996). The shark was reported measuring 600 cm in length (Anonymous, 1991b), and this measurement was later confirmed by Eric Licciardi, the son of Jean Licciardi (Vincent Maliet, pers. comm.). It was estimated to be 591 cm TOT by De Maddalena et al. (1991) on the basis of the photographic evidence. Both Séret (1996) and Fergusson (1996) erroneously reported this specimen as measuring 450 cm.

The report of a white shark, caught in January 1991, off Antibes that was transferred to the fishmarket in Rungis and reported by Touret (1992) most likely refers to the same shark caught off Sète, but with an erroneous capture location. Around 1993 some sportfishermen fishing for tuna sighted a white shark between Nice and Corsica, 40 miles offshore (Gérard Altman, pers. comm.; De Maddalena and Zuffa, 2008). Five years later, in 1998, an attack occurred off Cap d'Antibes. Two divers encountered a white shark that attacked one of them, biting his scuba tank. The fortunate diver did not suffer any injury, but only had scratches on the scuba tank from the teeth. The attack was reported in the local newspaper *Nice Matin* (Pierre Brocchi, pers. comm.; De Maddalena and Zuffa, 2008).

In the 2000s, there were more white shark reports from divers. On October 6, 2001, the dive center Abyss Explorers had a boat positioned on the site of the wreck of the *Miquelon*, near Marseille. The owner, Claude Wagner, hosted a group of nine advanced divers from Lyon and Toulouse, most of them at the dive instructor level. The wreck was located at a depth of 54 m, between Cap Croisette and Cap Caveaux. At 10:00 in the morning, the divers entered the water in groups. Wagner remained on the boat with the engine turned off. At the end of the dive, Sylvian and Sophie, who knew the dive site very well, were the first divers to surface. They were amazed, for they had never seen so many fish. After the first group boarded, Wagner steered the boat toward the second group of divers. The frantic divers immediately started to climb on the boat in a panic. One of the divers attempted to climb on the engine to get out of the water as soon as possible. When Wagner approached him to lend a hand, he suddenly understood. Wagner saw the huge shark swimming at the surface, no more than 20 m from the boat. Once the divers were on the deck, the panic was replaced with silence. Everyone was petrified. The shark swam slowly around the boat, but did not show any sign of aggression. It was longer than the boat, which measured 600 cm.

There were still two divers who needed to exit the water. Their decompression parachute appeared at the surface 30 m off the bow of the boat, while the shark continued circling the boat. Wagner realized that approaching the divers with the boat might result in bringing the shark to them. So he radioed for a fishing vessel to retrieve the divers, and he planned to motor in the opposite direction to lure the shark away from them. When the fishermen saw the shark, they were terrified and started to shout. Wagner started the engine of his boat, which caused the shark to immediately swim in the direction of the parachute. The huge shark began to circle the divers. One of the divers on the boat jumped into the water in order to help one of the decompressing divers get back on the boat. The last diver decided to remain in the water to complete the decompression. When the last diver finished his decompression and boarded the boat, the shark disappeared. The shark was estimated to measure 630–680 cm (De Maddalena and Herber, 2002; C. Wagner, pers. comm.).

On June 20, 2002, a white shark was encountered off Cap Ferrat. The dive master, Christian Gelpi, and two trainees were diving at the site called Caussiniere, located at the tip of Cap Ferrat, between Villefranche and Beaulieu. The depth of water was approximately 20 m, and the party were only 20 m from the shore. The weather was hot and sunny and the sea was calm. In the water, there was a large number of damselfishes *Chromis chromis*, European anchovies *Engraulis encrasicolus*, bogues *Boops boops*, European barracudas *Sphyraena sphyraena* and common dentex *Dentex dentex*. On the surface were many seagulls. At about 10:00, about five minutes into the dive, a white shark passed 10 m from them. The shark was near the bottom, at approximately 25 m in depth, and was swimming offshore. Gelpi saw the shark a brief 3 or 4 seconds and estimated it to measure about 400 cm. There was another diver spearfishing 30 m away who did not see the shark (De Maddalena and Zuffa, 2008; Christian Gelpi, pers. comm.).

Corsica

On September 14, 1984, a 527 cm white shark was caught near îlot des Moines, in the shallows of the Moines, 6 miles from Roccapina (northeast of Bouches de Bonifacio). The specimen was captured by Antoine-Jean Gianetti, Jean-Baptiste Gianetti, Thomas Duval and Antoine Duval, fishing with a trammel net aboard the vessel *Antoinette* belonging to the Gianetti brothers from Propriano, Corsica. The depth of capture was 80–90 m. The shark was entangled in the trammel by its caudal fin and was landed in Roccapina. It was reported as weighing approximately 1500 kg and its stomach contained a 60 kg dolphin. The shark was left on the beach during the night and thieves removed most of the teeth. Nevertheless, an anterior upper tooth was preserved by Roger Miniconi, of the conseil scientifique régional du patrimoine naturel d'Ajaccio. Photographic documentation of this specimen exists. The 500 cm length reported by an anonymous source (1984) is approximate, but the 850 kg weight mentioned by the same source may refer to the gutted specimen. In the same location, some months earlier, on May 7, 1984, a vessel owned by the Gianetti brothers sank after hitting an unidentified floating object (Anonymous, 1984; Miniconi, 1987, 1994; Roger Miniconi, pers. comm.).

On October 7, 1989, a shark was sighted swimming at the surface by Pietro Gambini, 1800 m from the mouth of the Solenzara River. However there is some doubt over the exact identity of the species (Fabrizio Serena, pers. comm.). During the night in February 1995, a shark was encountered by boat captain and professional fisherman Antoine Deriu and another witness 4 nautical miles (nm) off Capo di Feno. The shark passed below the boat and its size was estimated at about 600 cm by comparison with the size of the boat (Pierre–Henri Weber, pers. comm.).

In August 2008, an estimated 400–500 cm white shark was encountered by cetacean observer Pierre–Henri Weber while he was working for the company Corsica Mare Osservazione 0.6 nm offshore in the Golfe d'Ajaccio. The ocean conditions were good, with a clear sky and calm seas. Weber was alone on the boat when he spotted the shark, which swam very slowly and avoided the boat each time Weber tried to approach the animal. The shark was approached four times, and each time it was approached it dove and then emerged about 20 m from the boat. The encounter lasted at least 10 minutes (Pierre–Henri Weber, pers. comm.). A year later, in August 2009, an estimated 400–500 cm shark was encountered by seaman Christophe Recco and three other witnesses 2.5 nm offshore in the Golfe de Valinco. During the encounter, the shark breached, completely clearing the surface of the water. Film documentation of this shark exists, but it does not include the jump (Pierre–Henri Weber, pers. comm.). In December 2009, a large shark estimated at over 600 cm in length was encountered by boat captain Baptiste Bacchiolelli and another witness more than 50 nm offshore between Sagone and Toulon. The shark remained around the boat for approximately 45 minutes. Video documentation of this shark exists (Pierre–Henri Weber, pers. comm.). Most recently, in April 2010, an estimated 350–450 cm shark was encountered by boat captain and professional fisherman Antoine Deriu 0.2 nm offshore in the Golfe d'Ajaccio. The shark was also seen by approximately 10 other individuals that were aboard as passengers of Deriu's whalewatching boat (Pierre–Henri Weber, pers. comm.).

Monaco

The great white shark data bank has no records from Monaco.

Italy

There are three records from Italy that do not specify the exact location of the encounter. In 1881, a white shark was caught in an unknown location along the Italian coast. Its jaws are preserved at the Museo di Storia Naturale e della Strumentazione Scientifica in Modena, Italy (cat. no. PC-035/91) (De Maddalena, 1997, 2000c; Lawley, 1881). A few years after 1938, an attack on the boat of fisherman Maciotta occurred in an unknown location along the Italian coast. As a result, tooth fragments from the shark remained embedded in the wooden hull. The shark capsized the boat and ate the fish caught by the fisherman, including the fish boxes, but it did not touch Maciotta (Anonymous, 1938). In 1964, a white shark was caught in an unknown location along the Italian coast while it was feeding on a young sperm whale *Physeter macrocephalus* (Budker, 1971; Fergusson, 1996).

The Italian records with locations have been divided into the following areas: Ligurian coast, Sardinia, Tyrrhenian coast, Sicily, Ionian coast, and Adriatic coast.

Ligurian Coast

The earliest record from the Ligurian Sea dates from before 1846, when a 640 cm shark was reported caught off Santa Margherita Ligure (Fergusson, 1996; Sassi, 1846). Thirty years later, in 1876, a 446 cm male white shark was caught off Portofino. This specimen was donated to the Museo di Storia Naturale in Milano, Italy, with cat. no. 2142 (Tortonese, 1938) by Mr. Galletto. At the time, it was preserved but is currently missing. It is likely that this specimen was destroyed by the World War II bombings that devasted the museum in August 1943. Three years later, in 1879, a 315 cm female shark was caught off Viareggio. This shark is preserved as a taxidermied skin-mount at the Museo di Storia Naturale — Sezione di Zoologia "La Specola" of the University of Florence, in Florence, Italy (cat. no. 5983), and has a total length of 308 cm (De Maddalena, 1997, 2000c; Giglioli, 1880; Vanni, 1992). The following year, 1880, a shark was caught off La Spezia. The jaws are preserved at the Museo di Storia Naturale in Palermo, Italy. According to Mojetta *et al.*, this set of jaws should be the same jaws mentioned by Condorelli and Perrando (1909) (Mojetta *et al.*, 1997). These jaws are likely one of the five jaw sets preserved at the museum with cat. nos. An-108, An-115, An-128, An-145, and An-80 (De Maddalena, 2006b, 2007). In January 1883, a 430 cm TLn female shark was caught off Portofino. Among the items found in its stomach were dogs, cats, molluscs, a pair of pants, a pair of boots and pieces of canvas (Condorelli and Perrando, 1909). This specimen is preserved as a taxidermied skin-mount at the Museo di Storia Naturale e della Strumentazione Scientifica in Modena, Italy (cat. no. -045/91). According to Condorelli and Perrando (1909), the shark was caught around 1879–1880, but the label on the specimen bears the date of January 1883 (De Maddalena, 1999, 2000c).

More sharks were reported in the 1890s. On December 10, 1891, a female shark was caught off Monterosso al Mare. It was reported measuring 600 cm and weighing 600 kg gutted. The set of jaws was purchased from a market in Florence by the Museo di Storia Naturale — Sezione di Zoologia "La Specola" of the University of Florence, in Florence, Italy, where it is still on exhibit to the public in the section dedicated to the Selachians (cat. no. 6032) (De Maddalena, 1997, 1998, 2000c; Tortonese, 1956; Vanni, 1992). Five years later, on September 14, 1896, the carcass of an 18 m whale was found off Savona, 7 or 8 miles from Capo Vado. The source reported it was a blue whale *Balaenoptera musculus,* but it is likely that the identification was in error, since the blue whale is not a Mediterranean species and has never been reported from this area. It was most likely a fin whale *Balaenoptera physalus*, which is a common species in the Mediter-

This 430 cm female white shark caught in January 1883 off Portofino, Italy, is preserved as a taxidermied skin-mount at the Museo di Storia Naturale e della Strumentazione Scientifica of Modena, Italy (cat. no. -04991) (photograph by Alessandro De Maddalena, courtesy Museo di Storia Naturale e della Strumentazione Scientifica, Modena).

ranean Sea and can be mistaken for a blue whale. Some white sharks were observed around the cetacean carcass and were observed feeding on it (Parona, 1896). A month later, on October 19, 1896, the carcass of a 21 m whale was found floating 15 miles from Genova. Again, it was reported to be a blue whale *Balaenoptera musculus,* but once more, it was likely a fin whale *Balaenoptera physalus.* Some sharks were also observed around the cetacean carcass (Parona, 1896). The source was not clear about the species of these sharks, since he called them "pescicani verdoni (*Carcharodon*)," mixing *verdone,* which is one of the Italian common names of the blue shark *Prionace glauca,* with *Carcharodon,* which is the genus name of the great white shark.

Nine years after the turn of the century, in May or June of 1909, a shark was caught off Celle Ligure. It measured 350 cm in length and weighed 350 kg (Angelo Mojetta, pers. comm.). A few months later, in September that same year, a shark was caught off Sestri Ponente (Angelo Mojetta, pers. comm.). Five years later, again in September, an estimated 450 cm shark was caught off Portofino (Mojetta et al., 1997; Marco Zuffa, pers. comm.). The next report came over a decade later, on July 23, 1926, when an unprovoked fatal attack occurred 200 m off Varazze. At 11:00 a.m., a 20-year-old male, Augusto Cesellato, was swimming with his friend Luigi Baldi. Suddenly, Baldi heard Cesellato shouting and saw him disappear under the surface of the water. Then Baldi saw the shark. A boat that was passing nearby rescued Baldi, but Cesellato was nowhere to be found. The next day, his bathing suit and cap were found. The shark was estimated to measure 600–700 cm in length and the caudal fin upper lobe was said to protrude approximately 150 cm above the surface of the water (Anonymous, 1926a; Anonymous, 1926b).

Near the end of July in 1928, a provoked nonfatal attack occurred off Viareggio. Mr. Bagolini was aboard a small boat about 1000 m offshore when an unidentified shark approached the boat. The man was frightened and hit the shark with an oar, hoping to chase it away. However, the opposite occurred, as the shark attacked the boat, causing the man to fall into the sea. The scene was witnessed from a boat passing nearby that came to the rescue of Bagolini and chased the shark away (Gianturco, 1978). The source does not specify the shark species involved in this incident. But consid-

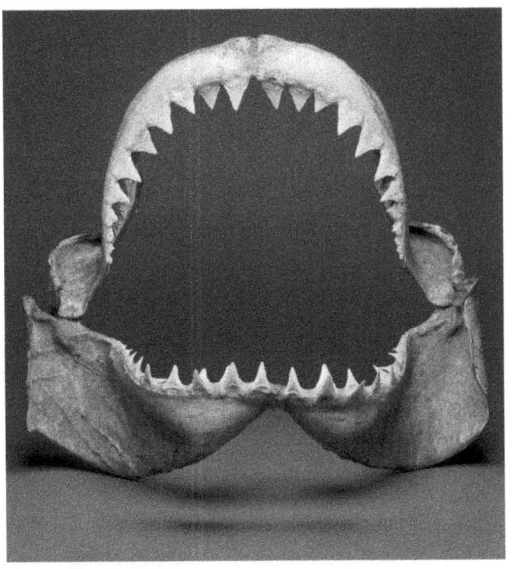

This set of jaws from an estimated 600 cm female white shark caught on December 10, 1891, off Monterosso al Mare, Italy, is preserved at the Museo di Storia Naturale — Sezione di Zoologia "La Specola" of the University of Florence, in Florence (cat. no. 6032) (photograph by Saulo Bambi/Museo di Storia Naturale di Firenze, courtesy Museo di Storia Naturale dell'Università di Firenze — Sezione di Zoologia "La Specola," Florence).

ering what happened, only a white shark (or less likely a large shortfin mako shark *Isurus oxyrinchus*) can possibly be the species responsible for an attack on the boat. A year later, in 1929, a 414 cm male shark was caught off Portofino. This specimen is preserved as a taxidermied skin-mount at the Museo di Storia Naturale in Genova, Italy (cat. no. C.E. 27517), and labeled as caught in the Golfo di Genova. This specimen is clearly the same one portrayed in the drawing that appears in Tortonese (1956). The TLn measured from the mount is 400 cm (De Maddalena, 1999, 2000c). This specimen has been recently restored by Enrico Borgo, the exhibit preparator and artist (Enrico Borgo, pers. comm.). On July 3, 1935, a shark was caught off Riva Trigoso, the set of jaws preserved at the Museo di Storia Naturale in Genova, Italy (cat. no. C.E.32695) (De Maddalena, 1999, 2000c).

Almost two decades later, in the month of June between 1953 and 1955, a juvenile shark measuring 80 cm in length was caught off Santa Margherita Ligure or Varazze (Angelo Mojetta, pers. comm.; Mojetta et al., 1997). However, some doubt exists over the exact identity of the species due to its small size. On May 16, 1954, a large female shark was caught in a small tuna trap off Camogli. The fishermen observed and reported that it was accompanied by a larger shark that swam around the area without entering

the tuna trap before disappearing offshore. The captured female was reported measuring 700 cm. However, Fergusson (1996) disputes the reported size, saying it was overestimated and should be closer to 520–550 cm TL based on the photographic evidence. However, the published photos cannot be used to make a valid estimate of the size of the specimen due to the position of the shark in the photograph. It was reported weighing 1400 kg and 2000 kg, depending on the source (Cappelletti, 1989b; Melegari, 1973; Tortonese, 1965).

On March 25, 1956, an attack on a boat by a white shark occurred off Genova. The shark was estimated to be 700 cm in length and attacked the boat near Punta Vagno (Mojetta et al., 1997). In the summer of 1961, a white shark was encountered by a diver off Camogli. The diver, F.K., was from Lugano, Switzerland, and was diving alone. He was spearfishing in the clear water a few hundred meters from the harbor of Camogli. He was accompanied by a boat that he had chartered occupied by some companions who were fishing. Suddenly, he saw an estimated 500 cm-long white shark swimming below him. The diver was petrified. The shark started to swim around him in circles, coming closer and closer. When the shark reached the same depth as the diver, it swam straight toward him. The diver closed his eyes and shouted, losing the mouthpiece of his regulator, and ascended rapidly to the surface, reaching the

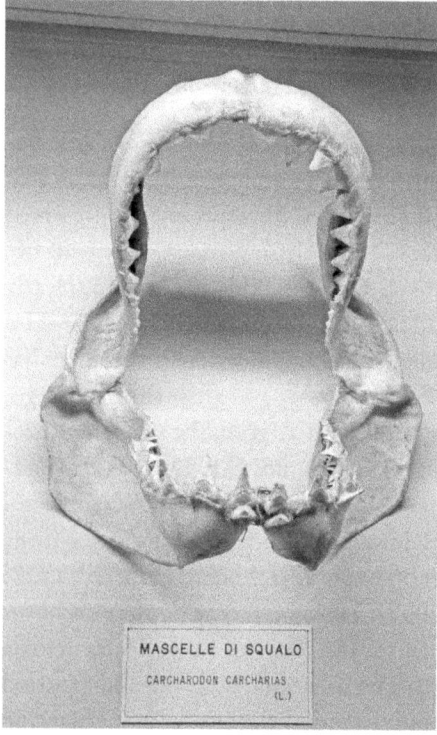

Top: This 475 cm white shark was caught in 1914 off Portofino, Italy. *Bottom:* This set of jaws from a white shark caught on July 3, 1935, off Riva Trigoso, Italy, is preserved at the Museo Civico di Storia Naturale "Giacomo Doria" in Genova, Italy (cat. no. C.E. 32695) (photograph by Stéphane Granzotto, courtesy Museo Civico di Storia Naturale "Giacomo Doria," Genova).

This 400 cm male white shark caught in 1929, in the Golfo di Genova, Italy, is preserved as a taxidermied skin-mount at the Museo Civico di Storia Naturale "Giacomo Doria" of Genova, Italy (cat. no. C.E. 27517) (photograph by Enrico Borgo, courtesy Museo Civico di Storia Naturale "Giacomo Doria," Genova).

safety of the boat (Roghi, 1962). Around 1970, a large specimen was caught in a tuna trap off Punta Chiappa (Albertarelli, 1990).

On June 15, 1983, a shark was encountered by a diver off Riomaggiore. The estimated 300 cm white shark harassed scuba diver Roberto Piaviali at a depth of 5 m. The diver did not suffer any injury (GSAF). On July 30, 1991, an attack occurred off Portofino. A 40-year-old woman, Ivana Iacaccia, was on a kayak only 20 m off the shore when an approximately 400 cm-long white shark approached the kayak. The shark hit the kayak, which caused the woman to fall into the water. The shark bit the kayak three times, leaving some tooth fragments embedded in the boat, but it did not attack the woman, who was rescued. The shark remained in the area approximately 15 minutes before it finally left. Based on the bite pattern and tooth fragments, Dott. Franco Cigala Fulgosi from the University of Parma identified the species responsible for the attack as a white shark (Fergusson, 1996; Graffione, 1991; Alberto Luca Recchi, pers. comm.).

Eight years later, on October 24, 1999, a large shark, likely a white, was sighted off Camogli. A coast guard helicopter from Sarzana was flying seven miles south of Punta Chiappa, searching for the wreckage of a small plane that had crashed into the sea. The depth of water was 570–580 m and the weather was rough. At 12:35, Captain Salvatore Cilona saw a large shark swimming at the surface. Characteristic of a white shark,

This reported 700 cm female white shark was caught on May 16, 1954, off Camogli, Italy (courtesy Editore Mario Bozzi, Genova).

the coloration was grey on the dorsal surfaces and white on the ventral surfaces. Cilona estimated the large animal to measure 600–700 cm in length (Salvatore Cilona, pers. comm.; Danker, 2001; De Maddalena, 2000a; Preve, 1999).

Sardinia

The earliest report from Sardinia was in 1777, when a shark was caught at an unknown location along the Sardinia coast (Angelo Mojetta, pers. comm.). More reports

This male white shark measuring longer than 400 cm was caught in April 1971 off Capo Testa, Sardinia, Italy (photograph by Renato Mariani-Costantini).

were recorded a century later, in the late 1800s. In the summer of 1879, a 405 cm-long white shark was caught in a tuna trap off Isola Piana. It was preserved as a taxidermied skin-mount at the Museo di Zoologia in Genova, Italy, but it is now missing and may have been destroyed in a flood that occurred in 1971. It is possible that one of two sets of jaws labeled as from an unknown location that are preserved at the Istituto di Zoologia in Genova may belong to this specimen (De Maddalena, 2000c; Parona, 1919; Marco Zuffa, pers. comm.). Almost three years later, on May 22, 1882, a shark entered a tuna trap, where it remained some days before it was finally caught. After a furious struggle, it was brought to the surface and killed with a hatchet. The total length was reported at over 300 cm (Parona, 1919).

Almost a century later, around 1970, an attack occurred near Mal di Ventre, in the Golfo di Oristano. A sportfisherman had caught some saddled seabreams *Oblada melanura*, from his 4–5 m long boat when a large white shark bit the keel and shook the boat, leaving some teeth embedded in it. When the boat was pulled out of the water, the teeth were extracted from the keel and examined by Prof. Mauro Cottiglia from the University of Cagliari, he identified the species as a white shark (Mauro Cottiglia, pers. comm.). A year later, in April 1971, a male shark measuring longer than 400 cm was caught in a tuna trap located 100 m off Capo Testa. Fergusson (1996) reported this case with a wrong location (Capo Falcone). The jaws of this specimen are preserved and displayed at the restaurant Le Bocche di Bonifacio in Capo Testa. A tooth from the specimen has been preserved by Renato Mariani-Costantini. Photographic documentation of this specimen exists (Renato Mariani-Costantini, pers. comm.; Marco

Zuffa, pers. comm.). Between the end of spring and early summer in 1975, a shark was caught nine miles west of Capo della Frasca, in the Golfo di Oristano. The shark was captured on the Secca della Frasca in a gill net (perhaps in a trammel net used to catch European spiny lobster *Palinurus elephas*). The depth of capture was 50–60 m. Prof. Mauro Cottiglia from the University of Cagliari briefly examined the shark and identified it as a white shark (Mauro Cottiglia, pers. comm.). The next year, on August 14, 1976, a male shark was caught in a tuna trap off Santa Caterina di Pittinurri. Depending on the source, it was reported to measure 400, 470 or 500 cm and weighed about 1100 kg. Its stomach contained the remains of sea turtles and some common dentex *Dentex dentex* (Fergusson, 1996; 1998c; Mojetta et al., 1997). In the same year, a shark was caught off Santa Teresa di Gallura. Photographic documentation of this specimen exists (Anonymous, 1976). In February, May or June of 1977, a 580 cm male was caught by Salvatore Murru in a tuna trap off Capo Testa. The jaws are preserved and displayed at the restaurant Le Bocche di Bonifacio in Capo Testa. Additionally, photographic documentation of this specimen exists (Mojetta et al., 1997; Marco Zuffa, pers. comm.). A year later, in April 1978, a small shark approximately 200 cm in length was caught in a small tuna trap 100 m off Santa Teresa di Gallura, near Capo Testa. Photographic documentation of this specimen also exists (Malatesta, 1987).

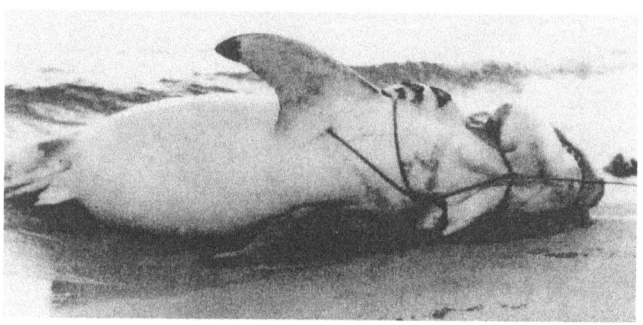

This 580 cm male white shark was caught in 1977 off Capo Testa, Sardinia, Italy (photograph courtesy Famiglia Murru).

Over a decade later, on August 24, 1991, a shark was sighted off Santa Teresa di Gallura (Angelo Mojetta, pers. comm.). In June 1997, a shark was encountered in the Golfo del Leone by the Italian navy. The weather was hot and the sea was flat. In the early afternoon around 13:30, Domenico Marcianò was serving aboard the Italian navy frigate *Scirocco*. A 50x50 cm box containing garbage from the kitchen, primarily pasta, was thrown into the sea and many seagulls swarmed the box, feeding on the garbage. Suddenly, a large white shark emerged and swallowed the box in a single bite. Two or three seagulls were also consumed, leaving the water red with blood. Marciano remembered the huge mouth, the snout and light grey coloration of the body. One hour prior to the encounter, some dolphins had been sighted in the area (Domenico Marcianò, pers. comm.). Two years later, in September 1999, a shark was encountered off Alghero. At 18:00, Tore Manca was surfing in shallow water only 1.5 m deep. He was in close proximity of a beach and several other people were in the water. Suddenly, Manca saw

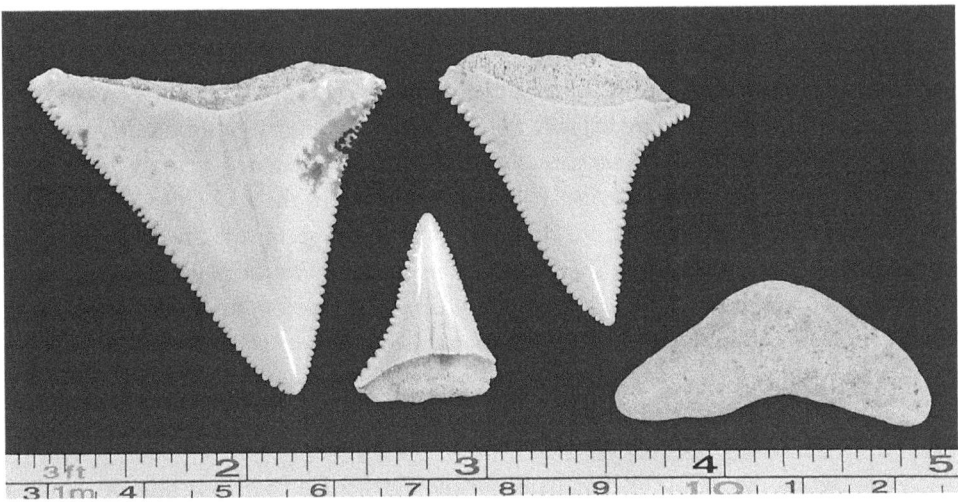

Four white shark tooth fragments found on September 1, 2009, and August 29, 2010, off Alghero, Sardinia, Italy (photograph by Marco Colombo).

a large white shark passing near him with its dorsal surface protruding from the surface. The animal disappeared as fast as it appeared, leaving the eyewitness awestruck. Manca estimated the animal to measure about 400 cm TL by comparing its length with the length of his surfboard (De Maddalena, 2000a; Tore Manca, pers. comm.).

After the turn of the century, a shark was encountered off Torre delle Stelle, also known as Capo Torre Finocchio, in the Golfo di Cagliari on August 29, 2001. Prof. Mauro Cottiglia from the University of Cagliari was fishing in calm seas from a 373 cm-long aluminum boat, approximately 300 m off Torre delle Stelle. The depth of the water was 30 m and the water temperature was 25°C. Around noon an estimated 400 cm-long white shark approached the boat and swam within 3–5 m from it, with the first dorsal fin cutting through the calm surface. It approached the boat from the stern, passed the boat and dove, all within a 10-second time frame. Some days prior to the incident, Cottiglia had seen a cetacean carcass, probably a sperm whale *Physeter macrocephalus*, in the area that could have attracted the shark. The carcass of a dolphin was also found a few days earlier with the caudal fin and part of the body missing from shark bites. Additionally, the fin of an unidentified animal had already been sighted at the surface, perhaps the dorsal fin of this white shark (Mauro Cottiglia, pers. comm.).

Six years later, on May 13, 2007, an estimated 300 cm-long white shark entered a tuna trap near Calavinagra, off Carloforte. The shark was seen by the divers inspecting the nets. It tore the nets in at least three places, allowing 200 bluefin tuna *Thunnus thynnus* to escape the trap. At the time, the chief of the tuna trap was Nicolo Puggioni (Froldi, 2007). In 2009, a large white shark was sighted by some fishermen many miles off Alghero, Sardinia, Italy. It was reported that a white shark, possibly the same individual, was sighted several times in the same area over the same period (Marco

Colombo, pers. comm.). On September 1, 2009, the crown of a white shark tooth was found by a snorkeler around 14:30 North of Alghero, Sardinia, Italy. The tooth lay on the seafloor at a depth of 3.5 m. Three other white-shark tooth fragments, two crowns with part of the root and a single root, were found by the snorkeler's son one year later, on August 29, 2010, in the exact same location. It was speculated that all tooth fragments were from the same individual that perhaps died in the area (Marco Colombo, pers. comm.). The teeth had white coloration and didn't seem to be fossils, appearing to be from a present-day white shark. Nevertheless, it cannot be completely excluded that they may have been only partially fossilized and are therefore subfossils, dating back to the Pliocene or Pleistocene (Henri Cappetta, pers. comm.).

Tyrrhenian Coast

The earliest record in the data bank for the Tyrrhenian coast was in 1666, when a great white shark was caught off Livorno. The head was acquired by the Medici in Florence, who gave it to Niels Stensen (Nicolas Steno) for dissection and study. By study the teeth of this specimen, Stensen discovered that the so called *glossopetrae*, or tongue stones, found in fossil beds were actually fossilized shark teeth (Stenone, 1667). Until then, fossilized shark teeth were believed to be tongues of serpents that Saint Paul had turned to stone while he was in Malta in 59 AD. On June 6, 1721, a shark was caught off Napoli. A few days before the capture, a person was collecting seafood in the waters off the beach called Ponte della Maddalena, in Napoli, when he was fatally attacked and eaten by a female great white shark. Following the attack, the shark was sighted several times in the area. The local fishermen decided to catch the animal and succeeded by placing a looped rope around its head. The shark measured 527 cm in length and weighed 1425 kg. Its stomach contained many fish and human remains, including a half cranium with hair, part of a vertebral column, and ribs. Very likely, these human parts were from the same person eaten a few days earlier (Ricciardi, 1721).

Several shark encounters were reported in the 1800s. In 1822, a small female shark was caught off Napoli. This 150 cm specimen is preserved as a taxidermied skin mount at the Senckenberg Forschungsinstitut und Naturmuseum in Frankfurt am Main, Germany (without cat. no.). This is the smallest taxidermied white shark preserved in Europe (De Maddalena, 2006b, 2007; Friedhelm Krupp, pers. comm.; Müller and Henle, 1838). Over a half century later, on May 7, 1883, a shark was caught off Scilla. It was harpooned at 3:00 p.m. in the back of the head by fishermen in search of swordfish *Xiphias gladius*. Still alive, the shark was towed by the boat for the rest of the day and through the entire night. Despite losing much blood from the wound, the shark continued to fight strenuously with the exhausted fishermen until the dawn of the following day, when a second fishing vessel assisted by tail roping the shark. The shark was landed on a beach called Spiaggia del Pezzo, where it finally died. This specimen weighed 1336 kg. The animal had remnants of two rusted harpoons embedded in its body that had been planted long before, since the wounds around the harpoons had healed. Its stomach contained many human bones, including a femur or a tibia and a cranium with hair. Human clothes were also found, including a pair of boots, three shoes, three socks, a wool bodice, and a pair of pants. A controversy arose over the ownership of some ancient Venetian golden coins found in the pants. This dispute had to be resolved in a court of law. One of the coins and three shark teeth were given to the minister state secretary of National Affairs (Anonymous, 1833).

In December 1876, a 1400 kg specimen was caught in the Canale di Piombino. The upper jaw perimeter was reported measuring 100 cm. A description of the jaws and an illustration portraying them were published in Lawley (1881). A decade later, in 1886, a huge white shark was caught in the Golfo di Baratti, off Piombino. The shark was trapped in a small tuna trap operated by Vittorio Canessa. Landing this large shark was very difficult, as the shark was dragged onto the beach of the "Casone" by two oxen. Its stomach contained an entire human corpse sewn into a sailcloth with cast-iron weights, clearly a seaman buried at sea. The seaman was later buried in the cemetery of Populonia. The specimen weighed over 2000 kg gutted and was reported measuring 800–1000 cm in length and having a mouth width of 150 cm. At the time of capture, the creature was erroneously identified as a porbeagle *Lamna nasus*. The shark carcass was sold at the market in Florence and the teeth were removed by the numerous onlookers that were present during the capture and landing of the shark. At that time, the news of this extraordinary catch caused a great sensation. More than a century after the capture, Vinicio Biagi succeeded in finding and examining two teeth from this shark (a photo of one tooth appears in Biagi, 1995). A third tooth measuring over 4 cm-long (according to Biagi, it belonged to the same shark), was found by a child in the summer of 1993 on the beach of the landing, near the Casone. Vinicio Biagi reported this record on the basis of a personal communication he received from the family Canessa (Vinicio Biagi, pers. comm.; Biagi, 1989, 1995; Mojetta et al., 1997). On June 5, 1898, a shark was caught off Portoferraio, Isola d'Elba. It was host to two different species of copepod parasites, *Echthrogaleus coleoptratus* and *Nemesis lamna var. sinuata*, that were attached at the dorsal fin insertion (Brian, 1906).

Many great white shark reports were recorded from the Tyrrhenian coast in the 1900s. On November 23, 1910, a white shark was observed feeding on a dying whale, probably a fin whale *Balaenoptera physalus*, off Marciana Marina, Isola d'Elba (Damiani, 1911; Mojetta et al., 1997). In 1924, a 1200 kg shark was caught between Scilla and Bagnara. It took eight hours for the fishermen to catch the shark (Gianturco, 1978). Some doubt exists over the exact identity of the species. The source did not mention the species, but based on the size it could only be a great white shark or a basking shark *Cetorhinus maximus*. In the same year, in mid June of 1924, a white shark was caught in the tuna trap Simeone off Procida. It was reported measuring about 600 cm in length and weighing about 1200 kg. However, it was estimated to be 507 cm TOT by De Maddalena et al. (2001) on the basis of the photographic evidence (Anonymous, 1924;

De Maddalena et al., 2001). In 1930, a 1400 kg specimen was caught in a tuna trap off Pizzo Calabro (Comune di Pizzo Calabro, 1991; Mojetta et al., 1997).

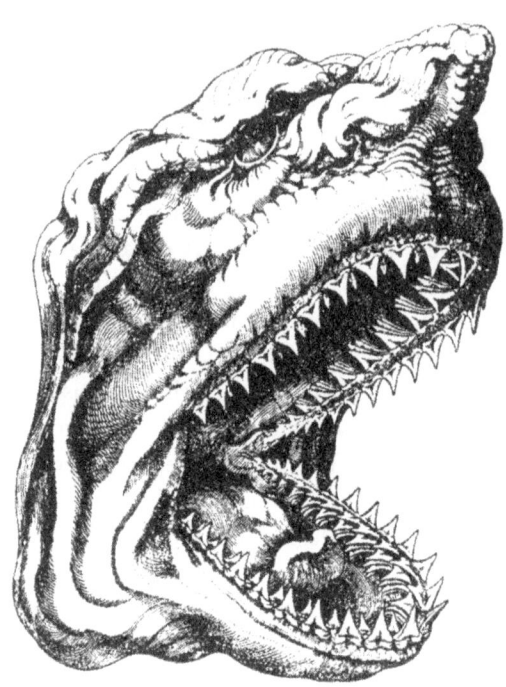

Head of a great white shark caught off Livorno in October 1666 and acquired by the Medici in Florence, who gave it to Niels Stensen (Nicolas Steno) for dissection and study (illustration from Stenone, 1667).

Three years later, in 1933, two sharks were caught in an unknown location of the Tyrrhenian Sea. The jaws were said to be preserved at the Museo di Storia Naturale in Genova, Italy, but the museum has only one set of jaws labeled from the Tyrrhenian Sea, while another set of jaws is from an unknown locality. One or both of these jaw sets could be those preserved at the Istituto di Zoologia in Genova (De Maddalena, 2000c; Mojetta et al., 1997). In the same time frame, around 1933 or 1934, a huge shark was caught by hook and line off Port'Ercole. It was reported measuring 800 cm in length. Storai et al. (2000) reported this record on the basis of a personal communication they received from Vinicio Biagi. Five years later, on August 12, 1938, a huge white shark, likely a female, was caught off Enfola, Isola d'Elba. The shark was trapped in a tuna trap belonging to the Ridi brothers. The stomach contained two dolphins and it was reported that the same shark was sighted several times in the area prior to capture, feeding on dolphins and tuna. At the time, it was reported measuring 510–600 cm in length and weighing 1800 kg. The longer measurement was confirmed based on an estimate of 597–613 cm TOT by De Maddalena et al. (2001) on the basis of the photographic evidence that still exists (De Maddalena, 1999; De Maddalena et al., 2001; Marco Zuffa, pers. comm.).

More reports were made during the 1940s and 1950s. In 1940, a shark may have been caught near Isola d'Elba in a tuna trap. This case is considered doubtful, since it may be the same capture that occurred on August 12, 1938, off Enfola, with a wrong date (Angelo Mojetta, pers. comm.). In 1949 or 1950, an attack occurred off San Vincenzo. The shark attacked the boat *Madonna della Libera* and bit the side of the boat (Vinicio Biagi, pers. comm.). There were no casualties from this attack. Another nonfatal boat attack occurred on the boat of fisherman Remo Adriani, 50 m off Sprizze, Isola d'Elba, on May 1, 1950. However, there is some question as to the exact identity of the species (Fabrizio Serena, pers. comm.). A few years later, in 1952 or 1953, an estimated 400–500 cm shark was sighted off Enfola, Isola d'Elba. This record was based on personal communication from Vinicio Biagi as reported by Storai et al. (2000). In September 1956, an attack on a diver occurred off Circeo, on the Secca del Faro. At 18:30, Goffredo Lombardo was diving in turbid waters, 1.5 miles offshore, where the depth of water was 16 m. Lombardo felt a strong blow to his upper back and turned to see the shark attacking him. The shark attacked the diver seven times, but the diver succeeded in fending off the predator with the spear of his speargun. Lombardo did not suffered any injury and only his SCUBA cylinders were scratched with teeth marks. A week later, Lombardo returned to the same site to catch the shark. He chummed the water and waited for the shark to take the baited hook. After a long time without any sign of the shark, Lombardo decided to dive and

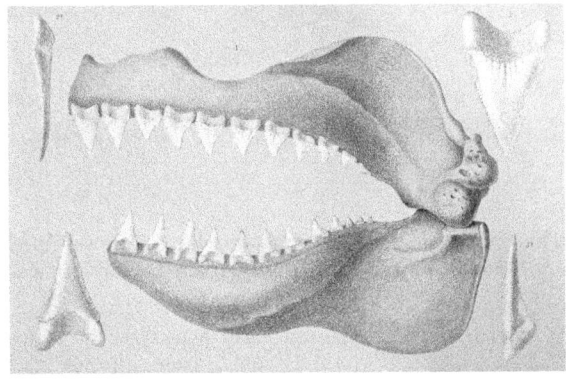

Top: This 150 cm newborn female white shark was caught in 1822 off Naples, Italy, and is preserved at the Senckenberg Forschungsinstitut und Naturmuseum of Frankfurt am Main, Germany (without cat. no.) (photograph by Sven Traenkner, courtesy Senckenberg Forschungsinstitut und Naturmuseum, Frankfurt am Main). *Bottom:* Jaws and teeth of a 1400 kg white shark caught in December 1876 in the Canale di Piombino (drawing by A. Manzella, from Lawley, 1881).

look for it. Once in the water, he found the shark, which had just taken the baited hook. As soon as the shark saw the diver, it charged toward him. Lombardo hid behind a rock until he could surface and climb back into the boat. Once aboard the boat, he brought the hooked shark to the surface and killed it with several rifle shots. The shark measured 420 cm in length and weighed over 600 kg. Photographic documentation of this specimen exists (Lombardo, 1960). Fergusson (1996) reported this case with a wrong date (September 1957).

Many white sharks were encountered off the Tyrrhenian coast in the 1960s. Around 1960, a possible attack occurred on three people (a mother, father and their six-year-old son) off Isola d'Elba (Hemingway and Devlin, 1965). However, this case is considered doubtful due to its resemblance to the case that occurred in 1909 off Augusta that was reported by Condorelli and Perrando (1909). In the same year, on August 18, 1960, a shark was caught 200 m off Rio Marina, near Miniera del Ginepro in water that was 50 m deep. The shark was captured in a trammel net operated by Gennaro Chiocca, Alfonso Chiocca, Giacomo Chiocca and Giuseppe Pinotti. On that day, the weather was good and the sea was calm. As the fishermen retrieved the net, they found it to be unusually heavy, and they understood why when a large white shark surfaced still alive. The shark was subsequently killed and landed. Its stomach contained two 15 kg dolphins. One of them was still intact and was sold the following day to a fishmonger. The shark was reported measuring between 600 and 700 cm in length and weighing over 1000 kg (Giuliano Chiocca, pers. comm.; Chiocca, 1990; De Maddalena, 1999).

This estimated 597–614 cm female white shark was caught on August 12, 1938, off Enfola, Isola d'Elba (courtesy Alberto Zanoli).

Another shark was encountered in the early summer of 1960, in the waters off Circeo. Maurizio Sarra, journalist and underwater photographer, was spearfishing in turbid waters on the Secca del Quadro. In order to spear a grouper, he entered a cleft. He then looked up and saw a large white shark passing in front of the cleft opening. Sarra estimated the animal to measure 700–800 cm. The diver remained motionless in the safety of the cleft, with the grouper next to him, and waited. The shark tarried for a while, swimming in wide circles, and eventually left. Sarra left the cleft and swam back to the safety of the boat. Some days before, a white shark had been sighted by many fishermen and seagoers in the same area (Carletti, 1973; Giudici and Fino, 1989).

In 1961 a white shark was sighted off Circeo (Carletti, 1973). Another sighting occurred in the summer of 1962 off Circeo (Carletti, 1973). At the end of the summer of 1962, a shark was encountered off Circeo by divers. Carlo and Raimondo Bucher saw a large shark, twice within a few weeks, while they were spearfishing. In both instances, the shark didn't show any aggressive behavior and fled when one of the divers attempted to touch it. While the two witnesses suggested it was a porbeagle, other reports recorded in the days that followed indicated that it was a great white shark. In the same time frame, a white shark was sighted by other seagoers in the area that may

This approximately 600 cm long male white shark was caught on August 18, 1960, off Rio Marina, Isola d'Elba, Italy (photograph by Gennaro Chiocca, courtesy Giuliano Chiocca).

have been the same shark. Near the end of August 1962, Carlo and Raimondo Bucher were once more spearfishing off Circeo when they saw, possibly, the same shark. The animal approached at very close range, swimming around them. When it got too close, one of the divers shot it straight in the gill slits with his speargun. The shark broke the line between the spear shaft and gun as it fled (Carletti, 1973; Giudici and Fino, 1989).

A few days later, at the end of August 1962, Dante Matacchione was diving off Circeo where he was attacked by a white shark, possibly the same specimen. Carlo and Raimondo Bucher observed the incident. Matacchione succeeded in chasing off the shark by poking it near the eye with his speargun. Attempts to catch the shark were unsuccessful (Carletti, 1973; Giudici and Fino, 1989). Weeks later, on September 2, 1962, a fatal attack occurred off Circeo. In the morning, Maurizio Sarra was spearfishing on the Secca del Quadro, where the depth of the water is 30 m. He had successfully speared many fish, which he had attached to his waist on a stringer. At 10:15 a motorboat approached the boat from which Sarra was diving to advise them that a white shark had been sighted in the area. When Sarra surfaced to load a freshly speared grouper onto the boat, he was advised about the presence of the shark in the water. Unbelievably, Sarra decided to continue diving. Five minutes later, at 10:20, he surfaced shouting and the water became stained red in color. The shark bit his legs, mauling them from the thigh to the ankle, the left leg being severely injured. The victim was immediately rescued and rushed to the hospital in Terracina within a half hour of the attack. Despite a surgery lasting four hours and requiring 250 stitches, the victim

Top: This estimated 400–500 cm white shark was encountered on September 19, 1989, off the mouth of the Arno River, Italy (photograph by Maurizio Bini e Tommaso Salone, courtesy Archivio Pesca in Mare). *Bottom:* This estimated 700 cm white shark was encountered on December 27, 1998, in the Golfo di Baratti, Italy (courtesy Capitaneria di Porto di Piombino).

did not survive, dying from irreversible shock (Carletti, 1973; Gianturco, 1978; Giudici and Fino, 1989; Marini, 1989).

On July 20, 1964, a large white shark was caught in the Canale di Montecristo. The shark was sighted while it was feeding on a fin whale *Balaenoptera physalus*. It was caught by Sirio Scotto, also called Sirio di Donna, from the fishing vessel *Antonio II*, south of Montecristo. Scotto caught the shark by putting a loop around its head. The creature was reported measuring over 900 cm and weighing 1900 kg. Its stomach contained a seagull, a small dolphin in two parts, a black wig, a broomstick and pieces of the whale. Pictures of this shark still exist (in

Rosselli, 1990). A few months later, on November 18, 1964, a male white shark and a female white shark were encountered off Circeo. Gino Felicioni and Cesare Polidori were spearfishing on the Secca del Quadro, at a depth of 30 m. The divers were temporarily separated when Polidori was approached by the female white shark, which he estimated to measure about 650 cm in length. The shark started to swim in decreasing circles around the diver. Polidori shot the shark with his speargun, but the spear tip did not penetrate the hard skin. Polidori then retreated to the surface with the shark following him. He was taken aboard the boat and rescued from the shark. During this time, Felicioni encountered the male shark and he too came back to the boat. The two sharks remained in the area for a while before leaving.

The next day, Felicioni and Polidori returned to the dive site in an attempt to catch the sharks. Using hook and line, they probably encountered one of the same sharks, which ripped out four hooks and remained near the boat from 7:00 in the morning to noon. The shark was accompanied by some pilot fishes *Naucrates ductor*. A few days later, Felicioni and Polidori succeeded in capturing the male white shark during the night of November 22, 1964. The shark was hooked and brought to the surface, where it was killed with rifle shots and a harpoon. The female shark approached the male and was harpooned at the base of the first dorsal fin, but she freed herself from the dart and fled. The boat headed toward Terracina, towing the male shark carcass. The female shark followed the boat at a distance and disappeared only when the boat approached the harbor. The male shark was measured at 320 cm and weighed 380 kg. This shark was photographed, documentation that still exists (Maltini, 1965).

On July 22, 1967, a huge white shark was caught in a net in the Canale dell'Isola del Giglio. It was reported measuring over 900 cm. Its stomach contained dolphins and large pelagic fish (Anonymous, 1967; Storai et al., 2000). In March of 1968, a shark was sighted off Civitavecchia. It was seen at 5:00 a.m. swimming at surface with the first dorsal fin and the caudal fin upper lobe protruding out of the water. The dorsal fin was estimated to be about 50 cm high. The shark followed a boat, owned by the company Tirrenia, which is a one-hour travel time from Civitavecchia. Some doubt exists over the exact identity of the species (Pirino and Usai, 1991). Two months later, on May 5, 1968, a white shark was caught in a net 100 m off Marciana Marina, Isola d'Elba. The depth of capture was 10 m (Fabrizio Serena, pers. comm.).

The 1970s was another decade filled with white shark reports. On an unknown date prior to 1971, a shark was caught off Argentario. It was captured in a trammel net on the Secca di Capo d'Omo. Its stomach contained two dolphins, a sea turtle and several kg of fish (Betti, 1971). On September 5, 1973, a shark was encountered by diver Dino Podestà three km off Vada, on the Secche di Vada, at a depth of 27 m. However, some doubt exists over the exact identity of the species (Fabrizio Serena, pers. comm.). Two years later, on September 15, 1975, Dino Podestà encountered another white shark on the Secche di Vada, at 15 m depth. Like the other shark, this shark was not positively identified as a white shark. (Fabrizio Serena, pers. comm.). On May 21, 1976, a white shark was sighted off Civitavecchia (Angelo Mojetta, pers. comm.). A year later, on October 26, 1977, a shark was reported on the Secche della Meloria, but it is unclear if it was captured. The reported length was 700 cm (Anonymous, 1979; Mojetta et al., 1997).

In 1978, a white shark was caught in a drift net, four miles off the coast in the proximity of Rende (the exact seaside location of capture was not reported, and Rende is 13 km from the sea). The drift net was fishing in water that was 500 m deep. According to the report, the shark measured approximately 500 cm and weighed 800 kg. Photographic documentation of this specimen exists (De Maddalena, 1997; Marco Zuffa, pers. comm.). In the same year, on September 16, a white shark, perhaps a female, was encountered in calm waters off Capo d'Anzio. At 11:00 a.m., Fabrizio Marini was diving in the 30 m-deep water and was surprised not to find the usually numerous groupers, seabreams and meagres. The lairs that usually were rich in fish were all empty on that day. Marini decided to collect some gorgonians instead of spearfishing. While he was dislodging a gorgonian from the bottom, he suddenly raised his head — just in time to see a shark ten m away coming straight at him. Terrified, the diver waved his speargun, hoping to keep the predator at bay. The shark attempted to attack the diver, but Marini succeeded in defending himself with the speargun. Then the diver shot the shark, and the animal fled with the spear embedded in its snout. Marini remained on the sea bottom for a while, hiding between two rocks, before venturing back to the surface. Marini estimated the specimen to measure about 500 cm (Marini, 1989).

The first recorded encounter of the 1980s in the Tyrrhenian Sea was in 1982, when a shark was reported to attack a boat off Biodola, Isola d'Elba. Giuseppe Campodonico was fishing when a white shark emerged from below and attacked the broad side of the boat, removing a portion of it. This case is considered doubtful (Perfetti, 1989; Marco Zuffa, pers. comm.). The second possible attack on a boat in 1982 occurred off Lacona, Isola d'Elba. The boat was owned by Giovanni Vuoso, who estimated the shark to measure 600 cm in length (Perfetti, 1989; Marco Zuffa, pers. comm.). In July of 1983 or 1984, two female sharks were caught off Baratti. On the date of capture, the weather was good and the sea was calm. The two sharks were caught offshore on a shoal, on a longline set the previous night. Both specimens were already dead when the longline was retrieved. A friend of author De Maddalena was onboard the boat at the time and saw the capture (at that time the friend was 12 or 13 years old). Della Rovere estimated the sharks to measure 350 cm and 300 cm, respectively. Della Rovere preserved two shark teeth, which are currently missing (Gianfranco Della Rovere, pers. comm.). Five months later, on December 3, 1984, an attack on a boat occurred 300 m off Marciana Marina, Isola d'Elba (Fabrizio Serena, pers. comm.). In 1987, another attack on a boat was reported to have occurred 300 m off Marciana Marina, Isola d'Elba. The shark bit the keel of the boat belonging to Aniello Mattera and Giorgio Allori (Perfetti, 1989; Marco Zuffa, pers. comm.). This case is considered doubtful.

In June 1987, a large shark was encountered off Scilla by fishermen. An estimated

150 kg swordfish *Xiphias gladius* was tied to a fishing vessel and was hanging in the water. A large shark engulfed the swordfish with a single bite, leaving only the head. No one aboard saw the shark (Giudici and Fino, 1989). On August 15, 1988, a shark was sighted 5.5 km off Procchio, Isola d'Elba. The shark was seen swimming at the surface by the commander of the research ship *Labromare*. Some doubt exists over the exact identity of the species (Fabrizio Serena, pers. comm.). Two days later, on August 17, 1988, an estimated 500 cm-long shark was sighted a half mile off Maratea (Angelo Mojetta, pers. comm.; Perfetti, 1989). Months later, on January 13, 1989, a shark was observed preying on a seagull in the Golfo di Baratti. However, this shark was not positively identified as a white shark (Fabrizio Serena, pers. comm.). Shortly thereafter, on February 2, 1989, an attack on a diver occurred in the Golfo di Baratti.

Luciano Costanzo was spearfishing near scoglio Stella, where the depth of water was 30 m. The diver was accompanied by his son, Gianluca Costanzo, and a friend, Paolo Bader, who were on the boat. At 10:25 a.m. a white shark estimated at over 600 cm in length appeared, with its large first dorsal fin cutting the surface of the water. The shark approached Luciano Costanzo, so Paolo Bader and Gianluca Costanzo started the boat engine, intending to go towards the animal and compel it to leave. But immediately the shark swam by in a wide circle and submerged. It then attacked the diver three times in just a few seconds. On the third attack, the shark bit the man on the chest and submerged, carrying the victim in its mouth. Only a few human remains and the divers' gear were found. One year later, the court of Livorno admitted a shark attack as the official cause of the death of Luciano Costanzo. This finally confirmed the account reported by the two eyewitnesses, vindicating them from unfounded rumors about the case spread by part of the Italian media (Albertarelli, 1990; Bertuccelli, 1989; Biagi, 1989; Cappelletti, 1989a; Giudici and Fino, 1989; Fabrizio Serena, pers. comm.).

More white sharks were encountered in 1989. On April 10, 1989, a shark was sighted by an officer from a Toremar company ferry in the Canale di Piombino. This was the same location where Luciano Costanzo had been attacked and killed by a white shark (Capuozzo, 1989; Angelo Mojetta, pers. comm.). On May 18, 1989, a large shark was again sighted in the Canale di Piombino (Anonymous, 1989b; Mojetta et al., 1997). It is possible that this case is the same as the previous one, but with a wrong date. On June 6, 1989, an unprovoked nonfatal attack occurred about 200 m off Marinella di Sarzana in rough seas. At 13:30 p.m., windsurfer Ezio Bocedi was bit on his upper right thigh by a white shark, estimated to be in excess of 300 cm, while Bocedi was leaning on his board. Bocedi was able to reach the shore on the board. Only then did he realize the severity of the wound and he used the surfboard leash to tie the injured thigh. He was immediately rescued and taken to the hospital, where he endured a surgery that lasted about two hours. The quadriceps femoris muscle was deeply cut and the wound required approximately 70 stitches. The judiciary bench launched an inquiry about the incident (Ezio Bocedi, pers. comm.; GSAF).

On September 19, 1989, a shark was encountered a half mile offshore, nine miles from the mouth of the Arno River. At 8:00 a.m., three sportfishermen, Maurizio Bini, Tommaso Salcone and Paolo Sbrana, were chumming in 65 m-deep water to attract tuna that were very numerous in the area at the time. Two other fishing boats were nearby. The anglers saw a large dorsal fin approaching the boat. The great white shark swam very close to Maurizio Bini's boat and was accompanied by some pilot fish *Naucrates ductor*. In order to keep the shark around the boat, the anglers continued chumming. The shark stayed around for some time in close proximity to the boats and when it left it, allowed the boats to follow it for a short distance. Photographs were taken of this shark that still exist. This specimen was estimated to measure between 400 and 500 cm in length (Conti, 1990).

White shark encounters started early in the 1990s. In 1990, a shark was reported off Livorno, but it is unclear if the specimen was sighted or captured (Angelo Mojetta, pers. comm.). Also in 1990, another shark was sighted off Baratti. However, some doubt exists over the exact identity of the species (Cappelletti, 1990). On May 25, 1990, a white shark was encountered 700 to 800 m off Favazzina. Fishermen Agatino Billé and Giacomo Lisciotto were fishing for swordfish *Xiphias gladius* from the felucca *Santa Rita*. The weather was good and the sea was calm. A great white shark was spotted and Lisciotto attempted to harpoon it, but the harpoon was deflected and broke. Lisciotto estimated the animal to measure over 1000 cm. Aricò (1990) identified this shark as a tiger shark *Galeocerdo cuvier*, but this was likely in error. Other shark sightings were recorded in the Stretto di Messina during the same time period, but without details (Antonio Celona, pers. comm.).

On May 15, 1990, a shark was sighted 300 m off Marciana Marina, Isola d'Elba. The exact identity of the species is unknown (Fabrizio Serena, pers. comm.). On an unknown date prior to 1991, a shark was sighted off Livorno pursuing a school of Atlantic bonito *Sarda sarda*. It was reported that the shark had followed the battleship *Andrea Doria* to Livorno (Pirino and Usai, 1991). On an unknown date prior to 1996, a white shark was reported off Livorno (Fergusson, 1996), but it is unclear if it was sighted or captured. On September 8, 1998, a white shark was encountered off Capo Vaticano. Fisherman Sisto Gurzì was on his boat towing an estimated 150 cm and 70 kg swordfish *Xiphias gladius*. Suddenly, a large white shark emerged headfirst from the water, severing the swordfish with a single bite and leaving only the head. Gurzì estimated the animal to measure approximately 600 cm in length. Gerlando Spagnolo reported this record on the basis of a personal communication he received from the fisherman, Sisto Gurzì (Anonymous, 1998; Gerlando Spagnolo, pers. comm.). Three months later, on December 27, 1998, a great white shark was sighted in the Golfo di Baratti. At 11:30 a.m., two sportfishermen, Roberto Cheli and Luca Antonio, were fishing near Scoglio Stella when a white shark appeared. It started to swim in circles around the boat, getting closer with each circle. It then dove but surfaced again shortly thereafter. The two anglers contacted the Piombino port captain's office, and a coast guard patrol boat was sent to the scene, arriving at 12:45 p.m. The shark continued to swim very close to the boats in the area, about 4–5 m from them most of the time. It remained on the spot until 1:30 p.m., then disappeared. It was estimated to be 700 cm TL by comparison with the length of the sportfishing boat, which measured about 600 cm. All the eyewitnesses agreed on this estimate. Photographic documentation of this specimen exists (De Sabata et al., 1999; Gasperetti, 1998). On an unknown date prior to 2000, a white shark was caught in an unknown locality of the Tyrrhenian Sea. The jaws are preserved at the Museo di Storia Naturale in Genova, Italy (cat. no. C.E.31916) (De Maddalena, 2000c, 2002).

The first report after the turn of the century from the Tyrrhenian Sea occurred on July 22, 2000, when a shark was almost caught near Isola del Giglio. Sportfishermen Alessandro Vitturini, Vito Cofrancesco, Giancarlo Graziani, Mario Marini, Salvatore Meli and another guest were fishing from the boat *Fishing Time*, a 12-m long Azimut 37 of which Vitturini was the skipper. They were 45 miles from Traiano, 22 miles off Argentario, on a shoal that was 470–500 m deep and surrounded by an area that was 1200 m deep. The weather was good and the sea was calm. They dropped five hooks baited with sardines positioned at a depth of 250 m using a lead weight and a surface float. Between 2:00 p.m. and 3:00 p.m., one of the floats started to move, and the anglers brought a shark to the surface. Vito Cofrancesco gaffed the shark and attempted to bring it into the boat headfirst. But the shark struggled, bending the 1 cm-thick gaff hook and freeing itself back into the water, where it immediately disappeared. This small shark was estimated to measure 200 cm in length and to weigh about 120 kg. Immediately after the shark left, the anglers saw the first dorsal fin of another white shark on the surface that was approximately the same size as the first shark (Vito Cofrancesco, pers. comm.; Porqueddu, 2000).

On February 11, 2001, a white shark was encountered on a shoal 38–39 miles southwest of Riva di Traiano. The depth of water was 570–600 m. The shark was sighted from the same boat as reported in the previous case, the *Fishing Time*. Onboard were Alessandro Vitturini, Vito Cofrancesco, Paolo Santi, Giancarlo Graziani and Luigi Signorelli. Sirocco had blown for the entire day but dimished in the afternoon. The sportfishermen hooked an estimated 20 kg grouper on the bottom, but when the fish was cranked up 100 m off the sea bottom, they felt a strong pull and the 1.4 mm nylon line was severed. A half hour later, at around 4:00 p.m., they saw the first dorsal fin and the caudal fin upper lobe of a great white shark emerge from the water. The anglers started to throw many fish into the water, including sardines, scorpionfish and squid. Every fish was eaten by the shark, which seized them as they sank two meters below the surface. Despite several attempts by the crew, the shark did not allow the boat to approach closer than 10–15 m. The animal stayed near the boat for about 45 minutes, but Vitturini decided to return to the harbor because he was worried by the excessive roll of the boat in the sloppy seas. The shark was estimated to measure 600–700 cm in length and photographic documentation of this specimen exists (Vito Cofrancesco, pers. comm.; Fanelli, 2001; Alessandro Vitturini, pers. comm.).

On October 7, 2002, a white shark was encountered on the Secche di Vada. Paolo Virnicchi, Paolo Spinelli, Franco Salomone and Alessandro Barlettani were diving from the boat *Maria Giustina*. They were at Ciglio di Terra, 2.68 miles southeast from the end of the pier belonging to the Solvay Society. The weather was good, the sea was calm and the water visibility was very good. The bottom depth of the water was 39 m. At 11:00 a.m., a great white shark arrived. Virnicchi had just left the water, while Salomone, Spinelli and Barlettani were still underwater decompressing from the deep dive. When they saw the shark, they remained at the decompression depth for some minutes, then got back to the boat. The shark was accompanied by about 10 pilot fish *Naucrates ductor*. As the predator swam in circles around the boat, the size of the shark was estimated to be 400–500 cm by comparison with the size of the boat, which measured 1400 cm. The shark departed around 11:40. On the sea bottom, there were many common two-banded seabream *Diplodus vulgaris* (Paolo Virnicchi, pers. comm.; Paolo Spinelli, pers. comm.; Alessandro Barlettani, pers. comm.). On June 10, 2006, a shark was sighted off Isola Gorgona. At 10:00 a.m. sportfisherman Vincenzo Giudice was 5 miles from Isola Gorgona, where the depth of the water was 250 m. Giudice saw the shark swimming at the surface for a few seconds and estimated it to measure between 500 and 700 cm (Vincenzo Giudice, pers. comm.). On the late afternoon of June 21, 2011, an estimated 400–430 cm female white shark was sighted swimming at the surface by researchers from the Museo di Storia Naturale — Sezione di Zoologia "La Specola" of the University of Florence, off Capraia. Filmed documentation of this specimen exists (Cecilia Volpi, pers. comm.).

Sicily

The earliest record in the data bank of a white shark in Sicily is in the 1800s. On an unknown date prior to 1864, an estimated 300 cm-long shark was caught in the Golfo di Catania (Gemmellaro, 1864). Around 1880, another 300 cm-long specimen was caught off Catania (Doderlein, 1881). The first record was of an attack that occurred around 1881, when a large white shark attacked a fishing boat in the Stretto di Messina (Doderlein, 1881). Mojetta et al. (1997) dated this case as occurring in 1862. In January 1889, a juvenile white shark was caught off Sferracavallo, near Capo Gallo. The jaws are preserved at the Museo di Storia Naturale in Palermo Italy. Morphometric measurements were provided by Riggio (1894). This jaw set is probably one of the five jaws preserved at the museum with cat. nos. An-108, An-115, An-128, An-145, and An-80 (De Maddalena, 2006b, 2007). In December 1892, a male white shark was caught off Capo Gallo and was landed in S. Lucia al Borgo. It was reported measuring over 400 cm and weighing 1500 kg. The specimen was on exhibit to the public in a warehouse located in Piazza S. Domenico in Palermo. It was not possible to preserve the entire animal, but at least the jaws were preserved in the ichthyological collections of the Museo di Storia Naturale in Palermo, Italy. Riggio (1894) compiled morphometric measurements of this shark. This jaw set is likely one of the five jaws preserved at the museum with cat. nos. An-108, An-115, An-128, An-145, and An-80 (De Maddalena, 2006b, 2007).

In August 1893, a white shark was caught off Pace by a fisherman with a harpoon used to catch swordfish. The shark was placed on exhibit to the public in a shack where people were allowed to view it for a fee. It measured 460 cm FOR and weighed about 1125 kg; the liver alone weighed 225 kg. The source reported some morphometric measurements (Facciolà, 1894). On an unknown date prior to 1894 a shark was caught in the Golfo di Catania (Riggio, 1894). In March 1894, a white shark was recorded in the Golfo di Catania, but it is unclear if this specimen was captured (Angelo Mojetta, pers. comm.). In the same month, on the 24th, a shark was caught near Isola delle Femine by some fishermen from Sferracavallo. It was reported to measure about 200 cm (Riggio, 1894). The jaws are preserved at the Museo di Storia Naturale in Palermo, Italy, and are likely one of the five jaws preserved at the museum with cat. nos. An-108, An-115, An-128, An-145, and An-80 (De Maddalena, 2006b, 2007).

In Sicily, a terrible earthquake took place on December 20, 1908. There was also a huge seaquake that caused enormous destruction in Messina. The large tsunami caused by the seaquake resulted in numerous deaths and disappearances. During this period, between the end of 1908 and the beginning of 1909, a large white shark was found stranded but still alive in Maregrosso, Messina. Its stomach contained the leg of a woman, cleanly cut (Berdar and Riccobono, 1986; Munthe, 1928). The victim undoubtedly was drowned by the tsunami. One month later, on January 26, 1909, another great white shark was caught near Capo Santa Croce, off Augusta, approximately 100 km south of Messina. Seven fishermen from Catania killed an estimated 100 kg dolphin. Immediately a female white shark surfaced and ate the cetacean in two parts. The fishermen proceeded to harpoon and kill the shark, which was then towed into the harbor of Catania. This female was measured at 450 cm and weighed 800 kg. It had the remains of at least three humans — a man, a woman and a six-year-old child — in its stomach. In this case, the victims undoubtedly were drowned by the tsunami. A detailed description and morphometric measurements of the shark, together with a description of the human remains found in its stomach, are provided by Condorelli and Perrando (1909).

On an unknown date prior to 1909, a 304 cm shark was caught in the Golfo di Catania (Condorelli and Perrando, 1909). Four years later, in 1913, a large white shark was caught off Torre Faro, near Messina. Its stomach contained an estimated 300 kg tuna that was swallowed in two parts. The jaws were preserved (Celona, 2002a). Over two decades later, on August 27, 1934, an attack occurred off Catania. A fishing vessel was hauling the net aboard the boat when an estimated 400 cm-long white shark approached the boat and bit a 150 kg fish that was entangled in the net. The fishermen tried to steal the fish from the shark, but the shark reacted violently, attacking the boat. Using harpoons, the fishermen fended off the shark, which eventually left injured (Giudici and Fino, 1989). Almost one year later, on May 23, 1935, a male shark was caught in the Stretto di Messina. It was sighted by fishermen from Torre Faro as it swam with other large sharks. This shark was captured, along with another shark of unknown species. The white shark's stomach contained two tuna that were 35–40 kg each. It was a large male and was examined by Prof. Concettina Scordia from the University of Messina, who reported the animal to measure over 500 cm (Scordia, 1935). Another male shark was caught on August 23, 1937, off Marzamemi. The motorboat *Sansone* was fishing in 35 m-deep water for European pilchards *Sardina pilchardus* when the shark arrived. The animal was captured with a harpoon only 500 m from the harbor. There were no contents in the stomach. It was reported to measure 530 cm in length and to weigh about 1200 kg. This shark was photographed and the documentation still exists. Celona (2002b) reported this record on the basis of a personal communication he received from the chief of the tuna trap in Portopalo.

In 1947 a huge shark was caught in a tuna trap off Scopello. It was estimated to measure 900–1000 cm and the weight was reported at 3590 kg. The teeth were said to be almost 7 cm-long, a tooth was preserved by Toti Lo Iacono. Photographic documentation of this shark exists, and although it does indeed show a huge shark, it unfortunately does not allow any estimation of the size (Maurizio Omodei, pers. comm.; G. Spagnolo, pers. comm.; Marco Zuffa, pers. comm.). Six years later, in May 1953, two large white sharks entered the tuna trap off Favignana. At the time, the chief of the tuna trap was Ernandes Flaminio. He attempted to catch the sharks by using whole lambs as bait, but he did not succeed. The sharks eventually freed themselves and fled the area (Giuseppe Guarrasi, pers. comm.).

On May 29, 1953, an estimated 550 cm white shark was trapped in a tuna trap off Favignana, Isole Egadi. The original source reported the shark as a male, but the photo clearly shows it was a female. The shark succeeded in breaking the nets and freeing itself. It then headed into the bay where the old tuna processing plant was located, likely attracted by the great amount of blood in the water. It remained there, swimming back and forth. In order to capture the shark, the fishermen tried to trap it by closing the small creek by means of nets, but they were unable to catch the animal. A little after noon, the military intervened and killed the shark with a machine gun. This large shark weighed over 1700 kg. Its liver weighed 201 kg and its stomach contained many large fish, including a 15 kg tuna, and garbage, including a 5 kg can. Oil was extracted from the liver, and the meat was reduced to fishmeal (Gioacchino Cataldo, pers. comm.; Giudici and Fino, 1989; Giuseppe Guarrasi, pers. comm.; Salvatore Spataro, pers. comm.).

Between 1960 and 1990, three white sharks entered the tuna trap off Favignana or Isola La Formica, Isole Egadi. Each shark succeeded in freeing itself and fled the area (Nitto Mineo, pers. comm.). In 1960, a large specimen was caught in a tuna trap off Capo Passero. It was reported measuring over 600 cm because when lifted with a block and tackle suspended from an architrave that was at a height of 600 cm a good portion of the shark was still touching the ground (Carletti, 1973). Photographic documentation of this specimen exists but is currently unavailable. On June 19, 1961, a large female shark was caught off Ganzirri. At noon, the shark was harpooned by Domenico Sorrenti from his boat. He and his four fishermen succeeded in capturing the animal after a one-hour struggle. Its stomach contained a large dolphin that had been swallowed in two parts. Sorrenti estimated the animal to measure about 600 cm and to weigh about 1500 kg. However, it was estimated to be 666 cm TOT by De Maddalena et al. (2001) on the basis of the photographic evidence that still exists (Celona, 2002a; De Maddalena et al., 2001). Two years later, in 1964, an estimated 600 cm great white shark was caught in a tuna trap off Isola la Formica, Isole Egadi. At that time the manager of the tuna trap was Mr. Laureato (Nitto Mineo, pers. comm.). However, according to another source, this date may be wrong, and the right date may be May 1972 (Giuseppe Guarrasi, pers. comm.).

In January 1965, a shark was caught off Marzamemi in 130 m-deep water on the Secca di Marzamemi, five miles offshore. The shark was entangled in a cable that was used to anchor pots used to fish caramote prawns *Penaeus kerathurus*. The cable became entangled around the caudal fin of the shark. As the shark struggled, the cable caused deep wounds. Other sharks bit the incapacitated shark on its caudal region, but the shark was still alive when the fishermen of the boat *Nella* found it. The animal was exhausted and gave up without a struggle. Accordino to the report, the shark was towed to Marzamemi and there it was measured at about 500 cm in length and weighed in at over 800 kg. The liver alone weighed 100 kg. The carcass was cut in pieces, to be used in part for human consumption and in part for bait. Its stomach contained bones and hairs, probably from a wolf or a dog (Marino, 1965).

Three months later, on March 9, 1965, a large shark was caught off Ganzirri, in the Stretto di Messina. The shark was first observed chasing a school of mullet *Mugil* sp. only a few meters from the shore. The water depth was very shallow, and the shark touched the bottom with the ventral surface of its body while its dorsal region was out of the water. As the shark moved towards the deeper offshore waters, it was followed by the boat of fisherman Nicola Donato. The shark was quickly overtaken and harpooned by Donato about 40 m from the shore. The harpoon penetrated through the posterior part of the trunk, at about midway between the first dorsal fin and the caudal fin. The wounded shark towed the boat for about four hours, crossing the Stretto di Messina three times before it finally died. Its stomach contained parts of loggerhead sea turtles *Caretta caretta*, remains of bony fish and other unidentified remains. Morphometric measurements of this specimen were taken by Professor Sebastiano Genovese of the University of Messina, who stated the length at 620 cm and the weight at 1200 kg. Celona et al. (2001) suggested that the 620 cm TL was obtained by measuring the shark along the curve of the body instead of in a straight line, so they revised the length to 560 cm TL for this specimen. The set of jaws from this shark is preserved in the Istituto di Idrobiologia of the University of Messina in Ganzirri (Celona et al., 2001). Many errors exist in the literature about the case. Mojetta et al. (1997) reported it with a wrong date (April 1). Berdar and Riccobono (1986) reported a pectoral fin length of 120 cm and a first dorsal fin height of 160 cm, which is clearly incorrect. An anonymous report (1965) stated the weight at 800 kg, but it most likely referred to the gutted and beheaded specimen. Fergusson et al. (2000) erroneously reported that the specimen had freshly ingested a 2 m-long ocean sunfish *Mola mola*. Fergusson et al. (2000) indicated the source of this data to be a personal communication that Ian K. Fergusson and Mark A. Marks received from Nicola Donato, but Donato never gave this wrong information to these two authors.

Also on March 9, 1965, a large shark was caught off Ganzirri. It was reported to measure over 700 cm and to weigh about 3000 kg. It was caught by Antonino Arena with the assistance of other fishermen. The shark struggled for nine hours before it finally succumbed (Berdar and Riccobono, 1986; Giudici and Fino, 1989; Mojetta et al. 1997). A few years later, in June 1967, an estimated 400–500 cm-long shark was encountered in the Stretto di Messina. The shark was harpooned by fishermen searching for swordfish, and it started to tow the boat toward the Sicilian coast, only a hundred meters away. Suddenly, when they were close to the shore, the shark dove and then surfaced under the boat, breaking its bottom and causing the boat to take on water. The fishermen realized that the boat was about to sink, so they were forced to jump into the water despite the presence of the harpooned shark. The men swam to the beach as fast as possible. The shark did not attack and all the men reached the shore unharmed (Giudici and Fino, 1989).

There were more white shark encounters reported off Sicily in the 1970s. Around 1970, a 400 cm shark was recorded off Siracusa (Fergusson, 1996; Touret, 1992), but it is unclear if it was captured. Another shark was encountered around 1970, about 1 km off Capo Rasocolmo. Fishermen aboard a 6 m-long boat had just finished catching squid when the shark hit the boat violently, causing one of the fisherman to fall into the sea. He was quickly retrieved while the shark bit the rudder. The shark then retreated and disappeared. It was unclear whether it had hit the boat deliberately or if it was a mistake (Antonio Celona, pers. comm.). On an unknown date prior to 1973, a shark was caught on the Banco del Mezzogiorno, 24 miles off Lampedusa, Isole Pelagie. It was entangled in the nets of a fishing vessel from Marsala (Carletti, 1973). On another unknown date prior to 1973, a white shark was caught 400 m from the harbor of Lampedusa, Isole Pelagie (Carletti, 1973). The following year, in the month of May 1974, a shark was caught in a tuna trap near Isola la Formica, Isole Egadi. The dead shark was found entangled in the net when the nets were checked in the morning. Nitto Mineo, the diver that regularly inspected the nets of the tuna trap, removed the teeth from the shark mouth while it was in the water. Mineo then distributed the teeth to the fishermen working at the tuna trap (the photo in Anonymous, 1974 that shows the mouth of the shark with its teeth must portray another specimen). Its stomach contained an entire goat *Capra hircus*, plastic bottles, plastic bags and possibly an entire dolphin weighing 80 kg. The carcass was exhibited to the public. The shark measured 620–640 cm and weighed 2400–2600 kg or approximately 1500 kg, depending on the source. Mineo specified that the specimen was measured several times by both fishermen and tourists. The liver alone weighed over 300 kg (Anonymous, 1974; Gioacchino Cataldo, pers. comm.; Fergusson, 1996; Giuseppe Guarrasi, pers. comm.; Nitto Mineo, pers.

comm.). This shark was estimated to be 594 cm TOT by De Maddalena et al. (2001) on the basis of the photographic evidence that still exists.

In August 1977 a 135 cm male was caught in an offshore net off Mazara del Vallo. Fergusson (1996) reported this record on the basis of a personal communication he received from Franco Cigala-Fulgosi. In 1978, a 1600 kg specimen was caught in the tuna trap off Favignana, Isole Egadi. The depth of capture was about 30 m. The diver who inspected the nets found the dead shark near the bottom. It had a plastic band, the type commonly used in packaging, tightly wrapped around its body (Gioacchino Cataldo, pers. comm.; De Maddalena, 2002; Salvatore Spataro, pers. comm.). In August 1978, a 220 cm shark was caught in a net off Mazara del Vallo. Fergusson (1996) reported this record on the basis of a personal communication he received from Franco Cigala-Fulgosi. Almost a year later, in June 1979, an estimated 200 cm-long shark was caught off Lampione, Isole Pelagie. Photographic documentation of this shark exists (Merlo, 1979).

The 1980s began with more shark encounters being reported. On April 24, 1980, two more sharks were caught in the tuna trap off Favignana, Isole Egadi. They were found by Nitto Mineo, the diver who inspected the nets of the tuna trap every day. The larger specimen, a female, was already dead and was immediately brought to land. The smaller one, a male, was still alive and struggling. It took two hours to tow the female to shore and land it. When the fishermen returned to the net to take the male, the shark was still alive. Since it had the caudal fin tangled up in the net, Mineo started to untangle it. Suddenly the shark arched its body to attack the man. Mineo saw the shark's head and put his hand on its snout. The shark straightened up then arched again, but Mineo distanced himself from the shark so it could not get closer than a meter from him. After that episode, the fishermen decided to wait until the following day to pull the male from the net. The following day (April 25, 1980), they found the shark dead and it was finally landed. The female measured 580 cm and weighed 1600 kg. Its stomach contained a partial skeleton of a swordfish *Xiphias gladius* and an estimated 20 kg dolphin in two parts. The male measured 540 cm. Photographic documentation of these specimens exist (Bruno, 1980; Gioacchino Cataldo, pers. comm.; De Maddalena, 2002; Nitto Mineo, pers. comm.).

In the month of May in one of the years between 1980 and 1982, a shark was caught in a tuna trap located over 5000 m off Bonagìa. The weather was quite good and the sea was fairly calm. The shark remained in the tuna trap together with 200 tuna, and 21 of these tuna became entangled in the attempt to escape the predator through the nets. The shark was caught in the morning near the surface at a depth of 3–4 m. The bottom depth was 46 m. The captured shark measured about 600 cm and weighed 1030 kg (De Maddalena, 2002; Ninni Ravazza, pers. comm.; Mommo Solina, pers. comm.).

On an unknown date prior to 1981, a pregnant female carrying six 40 cm-long embryos was caught off Sciacca. However, there is some doubt over the exact identity of the species (Di Milia, 1981). In August 1981, a 202 cm female was caught in a net 45 km southwest of Pantelleria. Fergusson (1996) reported this record on the basis of a personal communication he received from Franco Cigala-Fulgosi. This shark was photographed and the documentation still exists. Nine months later, in May 1982, a white shark was caught in a tuna trap off Scopello. The stomach contained the horns of a goat, a plastic container, a boat chain, an English navy boot, and half of an estimated 1.5 m-long striped dolphin *Stenella coeruleoalba*. It was estimated to measure 600–700 cm in length. The shark weighed 1300 kg, and the liver alone weighed 300 kg (Maurizio Omodei, pers. comm.). Over a year later, on August 10, 1983, a 155 cm-long male was caught in a net off Mazara del Vallo. This record was reported by Fergusson (1996) on the basis of a personal communication he received from Franco Cigala-Fulgosi. The next day, August 11, 1983, a 142 cm-long female was caught in a net off Mazara del Vallo. This record was also reported by Fergusson (1996), based on personal communication from Franco Cigala-Fulgosi. Around 1984, an estimated 500 cm white shark was caught in a tuna trap off Scopello (Gerlando Spagnolo, pers. comm.). Another small shark was caught on June 25, 1985, in a net off Mazara del Vallo. The shark measured 220 cm in length. Fergusson (1996) reported this record on the basis of a personal communication he received from Franco Cigala-Fulgosi.

In September 1985, a small shark was caught off Mazara del Vallo. The head was found at the fish market in Mazara del Vallo and the jaws were extracted and preserved. Vacchi and Serena (1997) estimated the animal to measure 138–158 cm on the basis of the jaw size. Sometime between 1985 and 1986, probably on a Sunday or a Monday in mid July, an attack occurred off Punta Secca. It was a very hot and sunny day, and the sea was flat calm. At 4:00 p.m., Neil Montoya, a 13-year-old boy, was snorkeling several hundred meters off Punta Secca. He was watching some small fish when he saw the large grey dorsal surface of a shark pass below him. He estimated the animal to measure 300–370 cm. Montoya was not scared of the shark and dove in an attempt to approach the animal, which was slowly moving away. The boy saw the shark disappear and was surfacing when he felt a strong hit to his left foot. He turned his head and saw the shark. Montoya kicked at the shark's snout and it veered away rapidly. The boy swam as fast as possible to the shore and reached it unharmed except for a bruise on his left foot. Even though there were others in the water, no one else saw the shark. Since Montoya had not suffered any significant harm, the case was not reported to the local authorities (Neil Montoya, pers. comm.). In March 1986, a 200 cm-long shark weighing 170 kg was caught in a net off Mazara del Vallo (Travaglini, 1986). About the same time in March 1986, a larger shark estimated at 500 cm was caught off Pantelleria (Mojetta et al., 1997).

On May 8, 1987, another shark was caught in the tuna trap off Favignana, Isole Egadi. On that same day, two other sharks, a male and a female, were caught in the tuna trap. The male succeeded in breaking the net and swam towards the diver that was inspecting the nets. The diver avoided the shark and the shark fled. The female did not escape and was caught. It measured 535 cm in length and had a huge girth of 440 cm. The shark weighed between 2000 and 2500 kg. A dolphin found in its stomach was already dead before it had been eaten, since it had a 3 m-long rope around its tail, indicating that it had been captured by fishermen. The cetacean weighed 150 kg, according to Gioacchino Cataldo (pers. comm.), but it was reported at 200 kg, according to Salvatore Spataro (pers. comm.). The fins of this shark are preserved at the Camperia in Favignana and photographic documentation of the specimen exists (Gioacchino Cataldo, pers. comm.; Curzi, 1987; Soletti, 1987; Salvatore Spataro, pers. comm.).

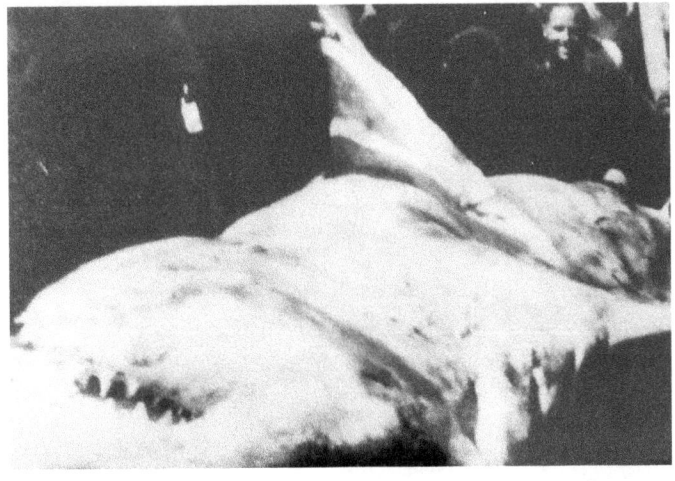

Top: This 530 cm male white shark was caught on August 23, 1937, off Marzamemi, Sicily, Italy (courtesy Società di Ricerca Necton). *Bottom:* This estimated 900–1000 cm white shark was caught in 1947 off Scopello, Sicily, Italy (photograph by Elio Bargione, courtesy of Gerlando Spagnolo).

A few months later, on July 31, 1987, the carcass of a 250 cm bottlenose dolphin *Tursiops truncatus* was found one km off Ganzirri; the carcass bore two white shark bites that were approximately 35 cm in diameter. The dolphin was bit on its lower surface closer to the tail. Some doubt exists over the exact identity of the shark species responsible (Centro Studi Cetacei, 1988). In 1989, a white shark was caught in the harbor of Isole Egadi, but the exact location is unknown (Fergusson, 1996). In the same year on June 23, a great white shark was sighted off Ganzirri by Antonio Celona from a boat fishing for swordfish *Xiphias gladius*. Celona estimated the animal to measure over 500 cm by comparison with the boat size (Antonio Celona, pers. comm.). Two months later, on August 13, three sharks, each estimated at 300 cm, were encountered by diver G. Bertin near Isola Galera, Isole Egadi (Fergusson, 1996; Mojetta et al., 1997).

The first encounter in the 1990s occurred around 1990, when an estimated 400 cm-long shark was caught off Palermo by Giuseppe Storniolo from the boat *Francesco Padre*. The shark's stomach contained the vertebral column of a swordfish *Xiphias gladius*, a dolphin that had been eaten in 3–4 parts, an albacore *Thunnus alalunga* eaten in two parts and some feathers. The shark was erroneously identified as a porbeagle *Lamna nasus*. Antonio Celona reported this record on the basis of a personal communication he received from the fisherman, Giuseppe Storniolo. On August 25, 1990, a male shark was caught by the fishing vessel *Santa Rosalia* 12 miles off Capo Granitola. The shark measured approximately 520 cm in length. When gutted, the stomach contained garbage, including metal cans. The carcass was landed in Mazara del Vallo. Since it was Sunday morning and the fish markets were closed, it was not possible to sell the shark so it was taken offshore and discarded at sea (Fergusson, 1996; Mojetta et al., 1997; Marco Zuffa, pers. comm.). A year later in June 1991, a 400 cm shark was caught off Mazara del Vallo.

Top: This estimated 666 cm female white shark was caught on June 19, 1961, off Ganzirri, Italy (photograph by D. Sorrenti, courtesy Società di Ricerca Necton). *Bottom:* This estimated 560 cm white shark was caught on March 9, 1965, off Ganzirri, Sicily, Italy (photograph by Antonino Donato, courtesy Società di Ricerca Necton).

This 620–640 cm white shark was caught in May 1974 near Isola Formica, Egadi, Italy (photograph by Antonino Rallo, courtesy Giuseppe Guarrasi).

Fergusson (1996) reported this record on the basis of a personal communication he received from Franco Cigala-Fulgosi, who examined the teeth or the jaws of the shark.

Between July 15 and 20, 1991, a female shark was encountered on the Banco di Pantelleria, where Riccardo Andreoli was spearfishing 22 miles offshore. The depth of water was about 50 m. On that day, the weather was calm and the sea was flat. Between 1:00 p.m. and 2:00 p.m., Andreoli had just seen some greater amberjacks *Seriola dumerili* and shot one weighing 6–7 kg. His friend saw a shark swimming at the surface and alerted the diver, who returned to the inflatable boat. Then they followed the shark with the boat. Andreoli succeeded in taking some photos of the shark poking its head out of the water. After a while, the shark dove, disappearing when a group of dolphins appeared at a distance. Andreoli estimated the animal to measure 490–500 cm. Photographs taken by Andreoli document this specimen (Riccardo Andreoli, pers. comm.).

This 540 cm male white shark was caught on April 24, 1980, off Favignana, Isole Egadi, Italy (courtesy Gioacchino Cataldo).

Fergusson (1996) reported this case with a wrong date (spring 1992) and a wrong size (530 cm).

In 1992 or 1993, a group of white sharks were sighted by some fishermen on the Banco di Mezzogiorno, off Lampedusa, Isole Pelagie. Fergusson (1996) reported this record based on personal communication he received from Antonio Testi. In May 1993 an estimated 180–200 kg tuna was severed in half by a single bite from a great white shark off Favignana, Isole Egadi (Mojetta et al., 1997). In the same month, a small shark measuring 140 cm in length was caught in a tuna trap off Favignana, Isole Egadi (Mojetta et al., 1997). However, this case is considered doubtful since according to Salvatore Spataro, ex-chief of the tuna trap, no such case occurred (Salvatore Spataro, pers. comm.). Maybe this reported specimen was not caught in the tuna trap. In the sum-

This 535 cm female white shark was caught on May 8, 1987, off Favignana, Isole Egadi, Italy (courtesy Gioacchino Cataldo).

mer of 1993, a small shark was caught in a net off Portopalo di Capo Passero (Mojetta et al., 1997). Later in November 1993, another small shark measuring 135 cm in length was caught. The location was off Mazara del Vallo and it was reported that the jaws were preserved. This record is based on personal communication that Fergusson (1996) received from Franco Cigala-Fulgosi.

In June 1995 or 1996, a shark was encountered near Strombolicchio, Isole Eolie. On that day, in the early afternoon, the sea was calm. At the time, the tuna were migrating in the area. The commander of Sarzana Coast Guard, Angelo Pistorio, was diving with some friends a depth of 40 m. They were taking photographs of gorgonians when an estimated 500 cm-long shark appeared. The predator showed no interest in the humans and disappeared after three minutes. Pistorio managed to get a photograph of the shark, but the photo seems to have been lost (De Maddalena, 2002; Angelo Pistorio, pers. comm.). In 1997, a 200 cm shark was caught in a net in the Canale di Sicilia. Storai et al. (2000) reported this record on the basis of a personal communication he received from Franco Cigala-Fulgosi. In the summer of 1997, many sightings of a white shark (possibly the same shark) by professional and sportfishermen were reported 2–3 miles off Marinella Selinunte and Capo Granitola, where the depth of water is about 40 m. The shark was estimated to measure about 300 cm in length (Capitaneria di Porto di Mazara del Vallo).

A year later, on August 14, 1998, a small shark measuring 150 cm in length was caught by Nicola Costa off Lampedusa, Isole Pelagie. Photographic documentation of this specimen exists (De Sabata et al., 1999). On October 19, 1999, a shark was encountered 20 miles south of Marsala. The sky was cloudy and the sea was calm. At 5:30 p.m., Giovanni Bosco was spearfishing on the Banco Chitarro at a depth of 47 m. He had seen some stingrays and was ready to spear the first fish of the day, a grouper, when he noticed the shark 10 m from him. The shark approached Bosco casually, passed above him and touched him lightly and then disappeared. Bosco estimated the animal to measure about 400–500 cm in length (Giovanni Bosco, pers. comm.). The next month, on November 2, 1999, a shark was encountered off Augusta, between Villa Marina and Faro Santa Croce. Skin diver and spearfishing champion Giovanni Zito was diving 700 m offshore, where the depth of the water was 25 m. At 1:00 p.m. Zito had just speared a grouper and noticed that the small fish nearby swam nervously, as if they were scared. Zito then spotted an estimated 500 cm white shark, which sighting compelled him to leave the water immediately (Nania, 1999).

More sharks were recorded in the data bank after the turn of the 21st century. On May 14, 2001, an estimated 100 kg specimen was encountered 200 m off Ganzirri. The weather was perfect, with a light breeze. The shark was sighted from a boat fishing for swordfish *Xiphias gladius* in water that was 35–40 m deep. Fisherman Agatino Billé was ready to throw the harpoon at the shark when the shark dove, disappearing into the depths (Antonio Celona, pers. comm.). The next year on June 29, 2002, an estimated 400 cm shark was sighted 400 m off Ganzirri in water that was 70–80 m deep. The weather was serene at noon when the shark was sighted by two boats fishing for swordfish *Xiphias gladius* (Antonio Celona, pers. comm.). On May 5, 2006, an estimated 350 cm bottlenose dolphin *Tursiops truncatus* was photographed near Lampedusa, Isole Pelagie, showing two fresh white shark bites on the dorsal region. Based on the photographed bite marks, the shark responsible for this attack was estimated to measure over 400 cm (Celona et al., 2006).

This 490–500 cm female white shark was encountered around July 15–20, 1991, on the Banco di Pantelleria (photograph by Riccardo Andreoli).

Two months later, on July 12, 2006, a small shark estimated at 150 cm in length was sighted from an inflatable boat 25 km off Lampedusa, Isole Pelagie, by Gabriella La Manna and other biologists from the Ricerche Delfini CTS of Lampedusa. Photographic documentation of this specimen exists (Ufficio Stampa CTS). On September 22, 2007, a shark was sighted on the Banco di Pantelleria, 24 miles off Pantelleria. It was a sunny day, with calm seas, and the water temperature was 24°C. Alberto Zaccagni was spearfishing and had encountered groupers, common dentex *Dentex dentex*, bullet tuna *Auxis rochei* and little tunny *Euthynnus alletteratus*, but he had not yet shot a fish. At 11:00 a.m., he was in waters 23 m deep, one meter above the bottom, when he encountered a great white shark swimming slowly near the bottom. The encounter lasted just a few seconds, as Zaccagni retreated and returned to the boat immediately (Alberto Zaccagni, pers. comm.).

Ionian Coast

Only five records in the data bank are from the Ionian Sea. On September 18, 1979, a huge male white shark was caught in the Golfo di Taranto, 6 miles from Gallipoli towards Torre S. Giovanni. It was entangled, with the caudal fin in the trammel net of Pompeo Alessandrelli set at a depth of 72 m. The net was completely destroyed by this capture. Marine biologist Pietro Parenzan asked his collaborator, researcher Roberto Basso, to examine the shark and to take measurements and photographs of it before it reached the Gallipoli harbor where he could examine the shark himself. Francesco Piccinno and Antonio Piccinno examined the specimen for the Stazione di Biologia Marina of the University of Lecce. The shark measured 620 cm in length, with the pectoral fins measuring 98 cm and the caudal fin upper lobe measuring 120 cm. A note written by Parenzan on a photo of the shark, which is still owned by Basso, confirms the total length at 620 cm. The same note included the weight of the shark, reported at 1700 kg. This weight differs by the one reported by Piccinno and Piccinno (1979), who reported that it weighed 2700 kg whole.

The stomach contained a pair of shoes, a washing machine drum, a doll, a ham bone, and one or more one-kg meat cans still sealed. Unfortunately, the fishermen, being very nervous about the damage to their net, did not allow the researchers to preserve any part of the specimen for study. The meat was declared unsuitable for human consumption. Photographic documentation of this specimen exists (Roberto Basso, pers. comm.; Piccinno and Piccinno, 1979). This may be the largest white shark ever examined by biologists who then confirmed its total length. The only problem is that no one seems to remember if the total length was measured in a straight line or over the curvature of the body. It has been said that the set of jaws was sold for two and half million Italian lire (1800 dollars) (Marco Zuffa, pers. comm.). According to Roberto Basso (pers. comm.), several teeth of the specimen were removed at the harbor prior to the removal of the jaws.

This 620 cm male white shark was caught on September 18, 1979, off Gallipoli, Italy (courtesy De Bellis).

On May 4, 1988, the carcass of a 310 cm bottlenose dolphin *Tursiops truncatus* was found in Bova Marina bearing a white-shark bite about 40 cm in diameter in the urogenital region (Centro Studi Cetacei, 1989; Marco Zuffa, pers. comm.). A year later, in mid June of 1989, a 550 cm white shark was caught in a drift net off Capo Spartivento. Its stomach contained two 120 cm dolphins (Centro Studi Cetacei, 1991). Around 2000, a white shark tooth was found by a skin diver off Punta Ristola, Italy. The tooth lay on the seafloor at a depth of 70 cm (Luigi De Giosa, pers. comm.). The tooth didn't seem to be a fossil, appearing to be from a present-day white shark. Nevertheless, it cannot be completely excluded that it may have been only partially fossilized and is therefore subfossil, dating back to the Pliocene or Pleistocene (Henri Cappetta, pers. comm.). On an unknown date, an estimated 450 cm male shark was caught off Marina di Salve. Photographic documentation of this shark exists. The photo does not allow an accurate estimation of the shark size, but author De Maddalena grossly estimated it to be around 450 cm.

Adriatic Coast

The earliest record in the data bank from the Adriatic Sea was on September 16, 1823, when a 490 cm TLn female shark was caught in an unknown location of the Adriatic Sea. This specimen is preserved as a skin-mount in the Museo Zoologico in

Padova (cat. no. P25E), making it the most ancient taxidermied white shark preserved in Italy (Canestrini, 1874; Condorelli and Perrando, 1909; De Maddalena, 2000c, 2002, 2006b, 2007; Paola Nicolosi, pers. comm.). In 1823, a shark was caught in an unknown location in Italian waters, probably in the Adriatic Sea. The set of jaws from this specimen is preserved in the Museo di Anatomia Comparata in Bologna, Italy (cat. no. Alessandrini 811) (De Maddalena, 2006b, 2007; Daniela Minelli, pers. comm.; Marco Zuffa, pers. comm.). In the spring of 1827, another shark was caught in an unknown location of the Adriatic Sea. It was exhibited at the Bologna fish market and the set of jaws is preserved at the Museo di Anatomia Comparata in Bologna, Italy (cat. no. Alessandrini 1216) (Daniela Minelli, pers. comm.; Marco Zuffa, pers. comm.; De Maddalena, 2000b, 2000c).

Over a decade later, in early February of 1839, a huge female white shark was either captured or stranded (the various sources differ on this point) in Civitanova Marche. It was reported to measure over 600 cm-long and to weigh 1814 kg. This specimen was shipped to Rome, where it was preserved at the University of Rome (Bonaparte, 1839; Condorelli and Perrando, 1909; Metaxà, 1839; Vinciguerra, 1885–1892). A detailed description, illustration and morphometric measurements of the shark were recorded by Bonaparte (1839). A 19th century illustration portrays this specimen attached to the ceiling of the Pontificium Romanum Archigymnasium. Vinciguerra (1885–1892) mentioned this specimen in a publication about the collections of the Museo Zoologico Universitario in Rome; and in the early 20th century, the specimen was mentioned by Condorelli and Perrando (1909). Recently, in the Museo Civico di Zoologia of Rome, the skin of this large shark was again preserved but was unfortunately lost or destroyed. In the Museo di Anatomia Comparata of Rome, part of the skeleton of this large shark, including the chondrocranium, the jaws and part of the vertebral column is preserved (cat. no. 111–95). From a morphometric analysis of the largest vertebra, the total length was calculated at 602 cm, making this the largest verified specimen preserved in an Italian museum (Ernesto Capanna, pers. comm.; De Maddalena, 1998).

On September 1, 1868, a fatal attack occurred off Trieste (Radovanovic, 1965; Soldo and Jardas, 2002). Between the years 1872 and 1905, the Imperial Maritime Austrian government issued three circulars offering a reward of up to 500 florins for every great white shark captured. These circulars also mentioned other shark species but primarily focused on *Carcharodon carcharias*. To obtain the monetary reward, fishermen were required to present their captured specimens to the Museo di Storia Naturale of Trieste for species identification. At the State Archives of Trieste, the orders of payment for these rewards still exist. Unfortunately, the species for which they were issued is not listed on the order in most cases. During these years, several sharks were reported as a result of this program. On April 19, 1872, a 300 cm-long shark was caught 8 miles off Grado (Brusina, 1888). A year later, in 1873, a 460 cm male was caught off Trieste (Doderlein, 1881; Graeffe, 1886). Four years later, on May 12, 1877, two sharks were caught in unknown locations along the Italian coast of the Adriatic Sea and were presented to the Central Maritime government to obtain the monetary reward (Ninni, 1912; Perugia, 1881). On November

5, 1879, a 250 cm-long shark was caught off Grado and was presented to the Central Maritime Government to obtain the reward money (Ninni, 1912; Perugia, 1881). In 1880, a 460 cm shark was caught in the Golfo di Trieste. It was preserved as a taxidermied skin-mount at the Museo di Storia Naturale in Venezia, Italy, but is currently missing (the set of jaws preserved at the museum with cat. no. 4860 Ist.V.S.L.A. 133 is not from this specimen) (Mojetta et al., 1997; Ninni, 1912). Before 1881 a 490 cm-long shark was caught in the Golfo di Venezia (Carus, 1893; Doderlein, 1881). On September 14, 1885, a 400 cm-long shark was caught off Santa Croce di Trieste (Brusina, 1888).

After the turn of the century in 1902, a 375 cm TLn male was caught off Trieste. This specimen is preserved as a taxidermied skin-mount at the Museo di Storia Naturale di Venezia, Italy (no.cat. 2039) (De Maddalena, 2000b, 2000c; Mizzan, 1994). In June 1908, a shark was caught by Stelio Candela in the Golfo di Trieste, which weighed 1400 kg (Arrassich, 1994). Almost four decades later, around 1945, an estimated 600 cm-long specimen was caught by Vittorio Pomante off Pescara (Cugini and De Maddalena, 2003). Almost two decades later, on July 7, 1961, at 2:00 p.m., in Riccione, a 21-year-old man, Manfred Gregor, was free diving and spearfishing when he was the victim of an unprovoked nonfatal attack by a white shark estimated to be 450 cm. The depth of water was 9 m. It was a cloudy day and the water was murky. Gregor was diving with two friends, Heinz Schmid and Herbert Russauer, and he was at the rear center of a reverse "V" formation. Suddenly, both front men turned around with their spearguns pointed towards Gregor. He tried to look back, but suddenly felt a flash of pain just as the shark bit his left foot and part of his swim fin. His companions shot the shark; one of the spear shafts penetrated its gills and the other pierced the center of the shark's body. The shark opened its mouth and Gregor pulled his foot from its mouth. The shark swam slowly into deeper water and Gregor was assisted to shore by his friends. His foot was severely bitten by the shark. When Gregor reached shore, his wound was bleeding heavily and first aid was rendered on the beach. Later that afternoon, he was attended by an Italian doctor and was transported to Austria, where he was hospitalized in Linz. Some sources cited 1963 as the date of this incident, but 1961 is the date given by Gregor himself (Baldridge, 1973; GSAF).

On June 7, 1978, a white shark was sighted in the Golfo di Venezia. Luigi Alberotanza and Luigi Cavaleri, two researchers from the Centro Nazionale delle Ricerche (CNR), were on the research platform *Acqua alta*, located 13 km off the Lido in waters 16 m deep. They were returning from a dive where they had cleaned the legs of the platform when they saw two dark fins on the surface. Believing it was a shark, they waited in hopes of getting a better view of the animal. Alberotanza tried to attract it by throwing a large steak in the water, but the fins disappeared. Some moments later, as the men headed inside the platform to take off their wetsuits, the platform was shaken by a powerful bump. The men clearly saw a great white shark swimming next to the platform. They estimated the shark's length, based on the known distance between the legs of the platform, to be about 500 cm-long. Luigi Cavaleri was able to take some photos of the shark. The shark disappeared, but later the remains of a bottlenose dol-

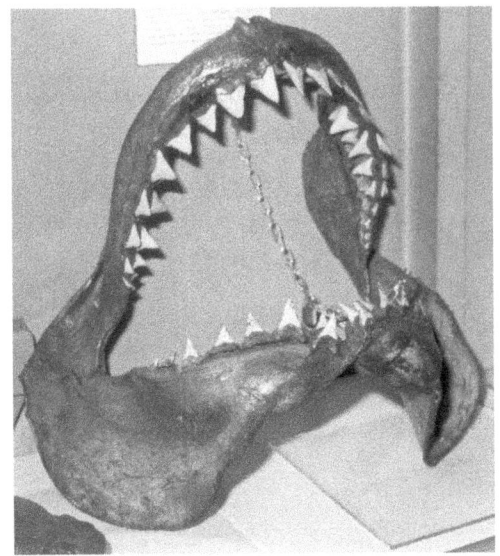

Left: This 375 cm male white shark caught in 1902 off Trieste, Italy, is preserved as a taxidermied skin-mount in the Museo Civico di Storia Naturale of Venice, Italy (cat. no. 2039) (courtesy Museo Civico di Storia Naturale, Venezia). *Right:* This set of jaws, from a white shark caught in 1823 in an unknown location in Italian waters, probably in the Adriatic Sea, is preserved at the Museo di Anatomia Comparata of Bologna (Dipartimento di Biologia Evoluzionistica Sperimentale, Alma Mater Studiorum, Università degli Studi di Bologna), Italy (cat. no. Alessandrini 811) (photograph by Marco Zuffa, courtesy Museo di Anatomia Comparata, Bologna).

phin *Tursiops truncatus* were found near the platform. Examination of the dolphin remains indicated that they were regurgitated by the shark, perhaps following its collision with the platform (Luigi Alberotanza, pers. com.; Albertarelli, 1990; Beltrame, 1983; Luigi Cavaleri, pers. com.). In Fergusson (1996), this incident is erroneously reported as happening in July 1977 and occurring in the Venice Lagoon. Also in June 1978, a few days after the previous case, a shark was sighted off Caorle. It was suggested that it was the same specimen, but there is no confirmation that this was the case (Luigi Alberotanza, pers. com.; De Maddalena, 2006b).

Almost a decade later, in late September 1986, there were several sightings of a large great white shark between Rimini and Pesaro (De Maddalena, 2000b; Gilioli, 1989; Giudici and Fino, 1989; Marini, 1989; Martelli, 1989). Most reports of the sightings of the shark estimate the length as being about 600 cm-long, but some estimates ranged as high as 800–900 cm. It may also be the same shark that attacked a fishing boat, and (possibly the same incident) snatched a whole crate of pilchards from the hand of a fisherman (Anonymous, 1986). This shark was first sighted on September 20 by the captain of the hydrofoil that travels along the Rimini–Yugoslavia route. Three days later, on September 23, the shark was sighted off Rimini near the oil-platform *Antonella*. On another occasion, it was seen 13 miles off Pesaro, near the oil-platform *Basil*. It was also

This part of the vertebral column from a female white shark in excess of 600 cm that was either captured or stranded in early February of 1839 in Civitanova Marche, Italy, is preserved at the Museo di Anatomia Comparata of Rome, Italy (cat. no. 111–95) (photograph by Alessandro De Maddalena, courtesy Museo di Anatomia Comparata of Rome, Italy).

reported that Roberto Bartomioli photographed and Marco Benelli filmed this shark, but it seems that unfortunately the pictures of the animal were never reproduced. Many anglers tried to capture the shark. Gabriele Bartoletti and Stefano Dragoni, on two separate occasions, succeeded in getting the shark to swallow a hooked bait, but they could not catch it. Several eyewitnesses described the shark as having a white coloration; perhaps they mistook a pale grey for white or possibly it was an albino specimen. Dubbed "Willy" by the fishermen of Rimini, this shark is attributed to several sightings that occurred from 1986 to 1989. It seems that the shark was seen near Pesaro between August and September 1986 and resighted during the same period in 1987 (Cardellini, 1987; De Maddalena, 2000b; Mojetta et al., 1997). It is unclear which characteristics of the shark allowed the sightings to be attributed to the same individual.

In May 1988, 28 miles off Numana, Fausto Fioretti sighted from his boat a great white shark that he estimated to be 450 cm-long, in water 85–90 m deep. The sighting occurred during a fishing tournament, so the shark was possibly attracted by chum. Photographic documentation of this specimen exists (De Maddalena, 1997, 1999, 2000b; Mario Marconi, pers. com.). Four months later, on September 9, 1988, there was a sighting near Porto Barricata. Some sportfishermen caught a large tuna, and a white shark approached the boat. The shark hit the chum sack, bit the boat engine several times and raised its head out of the water to investigate the boat and its occupants. Another boat arrived on the site and they tried to catch the shark. It bit the bait but wasn't hooked and then it left. It was estimated to measure over 550 cm. Photographic documentation of this specimen exists but it has not been possible to collect it (Mojetta et al., 1997, Marco Zuffa, pers. comm.). In September 1989, a shark was sighted off Pesaro. Estimated to measure over 500 cm, it was photographed. However, it has not been possible to obtain a copy. The record of this large shark was reported by Fergusson (1996) based on personal communication from Antonio Testi.

Several more sharks were reported from the decade of the 1990s. At

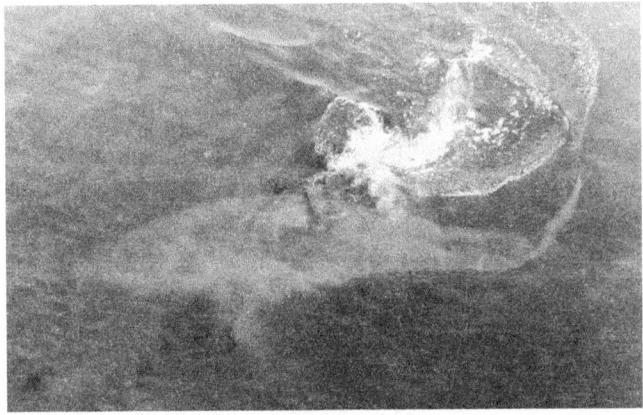

Top: This estimated 500 cm white shark was encountered on June 7, 1978, in the Golfo di Venezia, Italy (courtesy Luigi Cavaleri). *Bottom:* This estimated 450 cm white shark was encountered on May 1988 off Numana, Italy (photograph by Fausto Fioretti).

the end of August 1991, a specimen estimated over 600 cm was encountered between Pesaro and Cattolica. Armando Marzi was fishing for tuna with a friend when they got a call from a boat 1 mile distant from them stating that a huge shark was swimming below the other boat. Marzi reached the other boat and at the same time a third boat arrived on the spot. A tuna had just been caught and the shark was swimming close to the boats. Marzi saw its teeth and the marks they left on the wood of the boat. The shark trunk was said to be over 300 cm in width (Barone, 1999). On December 17, 1991, a shark was caught in a net off Ancona. The record of this small shark was reported by Fergusson (1996) based on personal communications from A. Testi and Marco Zuffa. It was reported to measure 210 cm and to weigh 180 kg. Before 1992, two sharks were caught in nets along the Italian coast of the Northern Adriatic Sea (Anonymous, 1992).

In mid March 1992, four or five young female sharks were caught off Termoli. Each shark measured approximately 230 cm and weighed about 200 kg (Anonymous, 1992; Fergusson, 1996).

In 1993, a shark was encountered 10 miles off Isole Tremiti. Fisherman Nino from Vieste saw the white shark upside down at the surface, with its ventral surfaces exposed and repeatedly gaping its lower jaw. It then started swimming and the fisherman tried to follow it, but the boat could not keep up with the shark, which disappeared. The fisherman estimated the animal to measure about 450 cm in length. Some doubt exists over the exact identity of the species (Gianluca Cugini, pers. comm.). Three years later, on September 15, 1996, a shark was sighted off Brindisi. Leonardo Leone De Castris was spearfishing 400–500 m off the harbor, accompanied by a friend that kept watch from a 470 cm inflatable boat. The depth of water was 10–12 m. The weather was calm and the sea was flat, but the water was turbid, with only 3–4 m of visibility. This area is rich in tuna *Thunnus* sp., mullet *Mugil* sp., greater amberjack *Seriola dumerili*, and common dolphinfish *Coryphaena hippurus*. An estimated 400–500 cm white shark approached the boat and raised its head out of the water. Photographic documentation of this specimen exists, but the quality is poor (Leonardo Leone De Castris, pers. comm.; De Maddalena and Herber, 2002; Marinazzo, 2001).

An estimated 500–600 cm white shark encountered on August 27, 1998, off Senigallia, Italy (photograph by Stefano Catalani).

On August 27, 1998, a large shark was encountered off Senigallia. At 3:00 p.m., some 22 miles off Senigallia in waters 72 m deep, a great white shark, estimated at 500–600 cm-long and about 1200 kg, came alongside the boat of Stefano Catalani. The 10-year-old son of Catalani was also with him on the boat. The angler had caught a common thresher shark *Alopias vulpinus*, which was tied to the side of the hull. The white shark circled the boat for about ten minutes, then bit the sack containing the chum and finally took a piece of the thresher shark carcass, leaving a 53 cm-wide bite. In order to protect their catch, the fishermen hauled the thresher shark carcass aboard the boat, but the great white shark remained close to the boat. After filming the shark for about half an hour, Catalani decided to leave. The footage filmed by Catalani was broadcasted by a major Italian television studio (Imarisio, 1998; Montefiori, 1998).

On September 26, 1999, another white shark was encountered by a fisherman

This estimated 600 cm white shark was encountered on September 26, 1999, off Giulianova, Italy (photograph by Elvio Mazzagufo, courtesy Capitaneria di Porto di Giulianova).

about 40 km off Giulianova. Around 3:00 p.m., Elvio Mazzafugo was fishing in waters 250 m deep in a spot called Fondaletto. He was on his fish charter boat *Coca Cola* with two guests. They had caught some tuna and tied them to the side of the boat when a great white shark approached the boat and ate a hooked tuna. Mazzagufo and one of his guests, Mario Galli, reported that the shark bumped the boat quite violently three times when the tuna blood fell into the water from the boat. Contrary to reports by the press, it seems inappropriate to mark this encounter as an attack. Mazzagufo shot the shark with a speargun, but the spear tip did not penetrate the shark's skin. He stated that he did not intend to harm the shark but just to mark it for future identification. The shark remained around the boat for almost two hours. Mazzagufo estimated this large animal to measure approximately 600 cm. Immediately before approaching the *Coca Cola*, the shark was sighted by fisherman Claudio Figorilli, who had his boat in the same proximity as the Mazzagufo boat and who confirmed the size of the shark. Photographic documentation of this shark exists (De Maddalena, 2000b; Fiaccarini, 1999a, 1999b; Graziosi, 1999). A few days later, between the end of September and early October 1999, another large white shark was encountered off Giulianova. Sportfisherman Antonio Bombini was on his boat *Biba 3* with four other persons. They had hooked a tuna at around 10:30 a.m. when an estimated 700 cm white shark approached the boat. The encounter lasted only about 15 seconds as the shark swam around the boat then left. This encounter was documented with video that was broadcasted by a major Italian television studio (Barone, 1999).

More sharks were reported after the turn of the 21st century. On April 29, 2000, a shark was encountered 1.5 miles off Faro di Piave Vecchia. The depth of water was 14 m and the weather was calm. An estimated 500 cm-long white shark approached the boat accompanied by some pilot fish *Naucrates ductor*, perhaps attracted by the chum. The shark remained near the boat for a while before leaving. One of the pilot fish was caught and given to the Museo Civico di Storia Naturale in Venezia, Italy. Luca Mizzan reported this record based on personal communication from Giovanni Sigovini (Luca Mizzan, pers. comm.). A year later, on April 13, 2001, a shark was encountered 1.5 miles off Torre Testa Rossa. At 4:00 p.m. Teddy Sciurti and Paolo D'Ambrosio were spearfishing in waters 21–24 m deep. The water was turbid and the visibility was 8 m. This area is abundant with common dentex *Dentex dentex*, greater amberjack *Seriola dumerili* and tuna, *Thunnus* sp. Within a time span of about five seconds, Sciurti saw a large white shark pass below him at 8 m distance, the visibility limit. Sciurti estimated the animal to measure 500–600 cm. D'Ambrosio did not see the shark, but he noticed that a school of Atlantic mackerel *Scomber scombrus* stayed tightly packed on the bottom and looked very nervous. The divers left the water immediately (De Maddalena and Herber, 2002; Teddy Sciurti, pers. comm.).

This estimated 700–800 cm white shark was encountered on September 9, 2002, off Porto San Giorgio (photograph by Glauco Micheli).

A few months later, on July 30, 2001, a shark was encountered 17 miles off Falconara. Tersilio and Giacomo Longhi were fishing for tuna from the boat *Triakis* in waters 65 m deep. It was a sunny day, with a light breeze and calm seas. Around noon, a tuna was hooked and a great white shark soon arrived. When the tuna was boated, the shark hit the boat with its snout. The shark swam circles around the boat for about 15 minutes before the anglers decided to leave. Giacomo Longhi estimated the animal to measure 550–600 cm. Photographic evidence of this shark exists (Giacomo Longhi, pers. comm.).

On September 9, 2002, a huge white shark was encountered 50 miles from Porto San Giorgio. Sport fishermen Glauco Micheli and Roberto Borracci were fishing from the boat Chat Noire. The depth of water was about 200 m. The weather was good and the sea was calm. At the time, the area was rich with bluefin tuna *Thunnus thynnus* and albacore *Thunnus alalunga*. At 7:45 a great white shark arrived swimming at surface with the first dorsal fin and the caudal fin protruding from the water. It was accompanied by 20–30 big pilot fish *Naucrates ductor*. Glauco Micheli estimated the animal to measure 700–800 cm. The shark left at 8:10. Photographic documentation of this specimen exists (Glauco Micheli, pers. comm.). On an unknown date prior to 2005 a white shark was sighted near the mouth of Po river. Photographic documentation of this specimen exists but it has not been possible to collect it (Marco Zuffa, pers. comm.).

Slovenia

There are only a few records of white shark encounters in the data bank from Slovenia. The earliest occurred on August 24, 1938, when an estimated 500 cm shark

This estimated 492–547 cm white shark was caught on October 22, 1963, in the Gulf of Pirano, Slovenia (courtesy Lovrenc Lipej).

carried away the net from the fishing vessel *San Giovanni* of Nicola Lubrano off Koper (Capodistria). It happened at 3:00 a.m., one mile from the Ospizio Marino. This case is recorded in a note that still exists at the State Archives of Trieste. On the basis of the shark's behavior and sheer size, it seems reasonable to infer that it was probably a great white shark (De Maddalena, 2000b). Around 1940, an attack on a boat occurred off Koper (Capodistria). As a result, a tooth fragment from the shark remained embedded in the wooden hull (De Maddalena, 2000b; Marco Zuffa pers. comm.). On October 22, 1963, a large white shark was caught off Salvore in the Gulf of Piran. It approached a fishing boat belonging to the Delamaris fish processing company while fishermen were retrieving their nets and was killed with 23 rifle shots. It was landed in Izola. The stomach contained an estimated 200 kg dolphin. One of the fishermen estimated the shark to measure approximately 600 cm-long and reported it to weigh about 1100 kg, but the local newspaper *Primorske novice* reported the shark to measure 400 cm and to weigh 700 kg (Anonymous, 1924, 1963; Bosnjak and Lipej, 1992–1993; De Maddalena, 2000b; De Maddalena et al., 2001; Lipej, 1993–1994). However, it was estimated to be 492–547 cm TOT by De Maddalena et al. (2001) on the basis of the photographic evidence.

Croatia

The data bank contains many records of white sharks encountered in Croatia. Unfortunately, most of the records contain very few details. The earliest report was on September 14, 1868, when a shark was caught off Jablanac. It is preserved in the Croatian Museum of Zagreb (Brusina, 1888). In the same year on December 16, a 460 cm specimen was caught off Sv. Gjuraj, near Senj. It is also preserved in the Croatian Museum of Zagreb (Brusina, 1888). There were several sharks reported in 1872. On April 16, 1872, a 490 cm shark was caught in the Preluka harbor. Its stomach contained the head and a leg of a man and a dolphin (Brusina, 1888; Hirtz, 1932; Soldo and Jardas, 2002). On May 12, 1872, a very small shark measuring 95 cm was caught 10 miles off Opuzen (Brusina, 1888). On May 12, 1872, a 237 cm shark was caught off Konao, Mljet (Brusina, 1888). On June 8, 1872, a 131 cm shark was caught in the Preluka harbor (Brusina, 1888). Days later, on June 16, 1872, a 146 cm shark was caught off Dugi Otok (Brusina, 1888). On July 25, 1872, a 260 cm shark was caught off Cavtat (Brusina, 1888). The last record from 1872 was on August 8, 1872; a 130 cm shark was caught off Rab (Brusina, 1888). Most of the white sharks reported in 1872 were small.

Five years later, on May 5, 1877, a 460 cm shark was caught off Ustrine, Cres (Brusina, 1888). Three days later, a similarly sized shark measuring 413 cm was caught off Sv. Martin, Cres (Brusina, 1888). On June 17, 1878, a 371 cm shark was caught off Osor, Cres. It was presented to the Imperial Central Maritime government to obtain the monetary reward (Ninni, 1912; Perugia, 1881). Less than two months later, on August 9, 1878, a shark was reported off Porec, but it is unclear if it was captured (Brusina, 1888, Soldo and Jardas, 2002). The following year, on May 21, 1879, a 382 cm shark was caught off Sv. Martin, Cres (Brusina, 1888). In June 1879, another shark may have been caught in the Kvarner, but this case may be the same as one of the following two cases that occurred in the same year (Fergusson, 1996; Graeffe, 1886; Tortonese, 1956). On July 23, 1879, a 402 or 445 cm shark was caught off Split. It was presented to the Imperial Central Maritime government to obtain the monetary reward (Ninni, 1912; Perugia, 1881). Three months later, on October 21, 1879, a 530 cm shark was caught off Ustrine, Cres. It was presented to the Imperial Central Maritime government to claim the reward money (Faber, 1883; Ninni, 1912; Perugia, 1881). In Fergusson (1996), this capture is dated September 1879; so maybe October 21 is the date of the presentation for the monetary reward cited above.

Many more sharks were reported in the 1880s and 1890s off Croatia. On April 22, 1881, a 380 cm white shark was caught off Rab (Brusina, 1888). Six months later, on October 16, 1881, another shark, measuring 405 cm in length, was caught off Rab (Brusina, 1888). On April 13, 1882, a 529 cm shark was caught off Sv. Martin, Cres (Brusina, 1888). The next year, on June 13, a 300 cm shark was caught off Vrboska, Hvar (Brusina, 1888). Three months later, on September 26, 1883, a 396 cm shark was caught off Rab (Brusina, 1888). A larger shark was caught on March 3, 1886, measuring 560 cm in length. It was caught off Korcula (Brusina, 1888). The next year on September 2, a 470 cm white shark was caught off KrK (Brusina, 1888). A similarly sized shark measuring 470 cm was caught in July off 1888 off Sv. Juraj. Its stomach contained a woman and a lamb (Hirtz, 1932; Soldo and Jardas, 2002). On April 26, 1891, a white shark was caught off Pag. It is preserved in the Natural History Museum of Zagreb, Croatia (Langhoffer, 1905; Soldo and Jardas, 2002). A year later, in September of 1892, a 450 cm shark was recorded off Bakarac, but it is unclear if it was captured (Hirtz, 1932; Soldo and Jardas, 2002). The following year, on February 19, a small male shark measuring 165 cm in length was caught off Zlarin (Katuric, 1893; Soldo and Jardas, 2002). On August 29, 1894, a 470 cm female shark was caught off Bakar. It is preserved at the Prirodoslovni Muzej in Rijeka (cat. no. PMR VP2) (De Maddalena, 2006b, 2007; Marcelo Kovačič, pers. comm.; Kovačič, 1998; Soldo and Jardas, 2002).

Reported encounters with white sharks from Croatia continued after the turn of the 20th century starting on July 15, 1901, when a 520 cm-long shark was recorded off Dubrovnik, but it is unclear if it was captured (Kosic, 1903; Soldo and Jardas, 2002). Two years later, on May 21, 1903, an estimated 600 cm shark was caught off Senj. The reported weight was 1200 kg and photographic documentation of this specimen exists. The capture is documented by a century-old postcard that author De Maddalena received from Klaus Riediger. On May 29, 1906, a 522 cm TLn female shark was caught in the Kvarner, probably off Bakarac. This specimen is preserved as a taxidermied skin-mount in the Museo di Storia Naturale in Trieste, Italy, and also has part of the vertebral column preserved (without cat. no.). This is the largest taxidermied great white shark preserved in Italy and one of the largest preserved in the world. This specimen is now under restoration (Andrea dall'Asta, pers. comm.; De Maddalena, 1999, 2000c).

In January of 1908, near Medola, some sharks approached a boat filled with young women. One of the sharks may have attacked the boat, because Milena Scambelli fell suddenly into the sea. A shark bit her leg, severing it. Scambelli was rescued and taken to a hospital, where she died (Anonymous, 1908; Marco Zuffa, pers. comm.). Whatever caused the attack, as well as the identity of the species, is uncertain. It seems interesting that the women specified that the sharks "jumped around the boat" (Anonymous, 1908; Marco Zuffa, pers. comm.). On May 19, 1908, a small shark measuring 170 cm in length was caught off Stadival by fishermen Simeone Armanini and Simeone Franceschini. It was incorrectly identified as a shortfin mako shark *Isurus oxyrinchus*. The case is documented by an order of payment for the reward (Marco Zuffa, pers. comm.). A year later, in October 1909, a great white shark, possibly measuring 550 cm, was caught in a tuna trap off Kraljevica (Angelo Mojetta, pers. comm.). In Mojetta et al. (1997), a different location (Rijeka) is reported for this capture.

Over a decade passed with no reports. The next report was on February 2, 1920, when a 525 cm shark was caught off Dugi Otok, Kornati. Its stomach contained a dolphin and the shark was reported to weigh 1300 kg. Soldo and Jardas (2002) reported this record on the basis of an unpublished manuscript by Kisic. Six years later, in March 1926, a 500 cm white shark was recorded off Ugljan, but it is unclear if it was captured. However, it was reported weighing 700 kg (Morovic, 1950; Soldo and Jardas, 2002). In August of 1926, a white shark was caught off Lumbarda. It was reported to measure 400 cm and to weight 500 kg. Its stomach contained human remains (Hirtz, 1932; Morovic, 1950; Soldo and Jardas, 2002). Two months later, in October, a large white shark was caught in a net off Lumbarda. It was reported to measure 600 cm in length and to weigh 1800 kg (Morovic, 1950, 1973; Soldo and Jardas, 2002). In 1927, another large shark was caught off Rovinj. It was reported measuring about 600 cm and weighing 1000 kg. Its stomach contained several inedible items. Photographic documentation of this specimen exists (De Maddalena, 2000b).

There were more white shark reports in the 1930s, including two fatal attacks. In 1931, a small female shark 150 cm in length was recorded off Rogoznica, but it is unclear if the shark was captured (Anonymous, 1931; Soldo and Jardas, 2002). On August 21, 1934, in Susak, an attack occurred on an 18-year-old girl, Agnes Novak, who was swimming near a tuna trap. At this location, there was a swimming area with an anti-shark net, but Novak entered the water outside this net. Eyewitnesses from a fishing boat heard the woman cry out and saw a large great white shark bite Novak's abdomen and carry her underwater (Giudici and Fino, 1989). In the days that followed the attack on Agnes Novak, there were many sightings of sharks. At least two sharks were sighted on August 23 and August 30 near Rijeka. On August 23, 1934, an estimated 600 cm shark was seen by some soldiers as it swam near a torpedo factory. Possibly the same shark

Top: This approximately 600 cm white shark was caught on May 21, 1903, off Senj, Croatia. *Bottom:* This 522 cm female white shark caught on May 29, 1906, in Kvarner, probably off Bakarac, Croatia, is preserved as a taxidermied skin-mount at the Museo Civico di Storia Naturale of Trieste, Italy (without cat. no.) (photograph by Alessandro De Maddalena, courtesy Museo Civico di Storia Naturale, Trieste).

This 460 cm female white shark was caught on December 10, 1955, off Senj, Sveti Juraj, Croatia, and is preserved as a taxidermied skin-mount at the Zemaljski muzej Bosne i Hercegovine, in Sarajevo, Bosnia and Herzegovina (without cat. no.) (courtesy Zemaljski muzej Bosne i Hercegovine, Sarajevo).

was sighted later that afternoon by some fishermen; it was seen swimming toward shore off Diga Cagno. On August 30 two large sharks were reported between Punta Baro and Diga Cagno. An hour later, a shark estimated to be more than 700 cm-long was swimming towards the channel of Labin where it was seen by some fishermen (Giudici and Fino, 1989).

Nine days after the fatal attack on Novak, another swimmer was fatally attacked by an estimated 600 cm white shark. This attack occurred off Reotore. On August 30, 1934, the *New York Evening Sun* carried a story datelined Rijeka. The strange story told of a young girl, Zorica Princ (probably misspelled as Zorca Prince and later as Zorca Prinz), who paid with her life because she did not believe in dreams. Her mother had pleaded with her in a letter not to swim far from shore, for she had dreamed that Zorca would fall victim to a shark. Exclaiming to her friends, "I don't believe in dreams," the young student, a strong swimmer, swam toward a fishing boat far out in the sea off Reotore. The fishermen heard a shriek and went to help the girl only to find nothing but bloodstained water. They reported that a shark had been seen earlier swimming around the edge of their nets. The authenticity of this case was regarded as questionable for a long time because of efforts to make the public believe the case was a fabrication organized by the *New York Evening Sun* (Baldridge, 1974). The authenticity was further challenged by a report with a dateline of Belgrade, 1 September 1934, from an

2—Records of Great White Sharks from the Mediterranean Sea

This white shark was estimated at approximately 200 cm and was caught in 1968 near Rava, Croatia (courtesy Lobert Simičić photograph archive/Robert Lončarić).

article in another magazine that read, "It is reported from Kraljevica in the Alvala District that the news published by certain foreign papers according to which a young Yugoslavian girl, Miss Prinz, was attacked and eaten by a shark off the Italian coast [at that time the attack location was still Italian territory], is without foundation. Miss Prinz is actually at her parents' home in Ljubljana and intends to spend the coming month taking examinations for admission to the university of this town" (Baldridge, 1974). In fact, the fatal attack actually occurred (Lovrenc Lipej, pers. comm.).

Three days after the second attack, on September 2, 1934, an enormous shark was caught in Kralijevica. Depending on the source, it was reported to measure 775 cm or over 700 cm-long and to weigh 1100 kg or 2000 kg. Examination of its stomach contents did not indicate that it was responsible for the attacks on the humans that had occurred in the previous days (Giudici and Fino, 1989; Morovic, 1973; Soldo and Jardas, 2002). Five days later, on September 7, 1934, a large shark was caught in a tuna trap off Moscenicka Draga. Depending on the source, it was reported to measure 600 cm or about 500 cm and to weigh 1000 kg or about 800 kg (Anonymous, 1934; Giudici and Fino, 1989; Soldo and Jardas, 2002). The same day, only a few hours after this capture, a shark longer than 6 m was seen near Martinschizza and an hour later the same shark was likely sighted, near a fishing boat, eating a small board of cork (Giudici and Fino, 1989). Almost a year later, on July 20, 1935, another large shark was caught in a tuna trap off Lukovo. It was reported measuring 600 cm and weighing 2500 kg (Anonymous, 1935; Morovic, 1950; Soldo and Jardas, 2002).

A decade passed after these attacks with no reports. Then, in the summer of 1946, a shark was caught off Bakarac. Its stomach contained a 10 kg domestic pig *Sus scrofa domestica* (Jolic, 1988; Soldo and Jardas, 2002). Four years later, around 1950, a great white shark was sighted in the harbor of Rovinj. The shark hit the hull of a boat with the dorsal part of its body. The shark was harpooned by sailor Peppo, but it escaped. Daniel Abed-Navandi reported this record on the basis of a personal communication he received from Prof. R. Riedl and Prof. J. Ott. (Daniel Abed-Navandi, pers. comm.). In August of 1950, a huge white shark, estimated to be 7–8 m long, was sighted off Primosten, scavenging on a calf *Bos taurus* carcass (Anonymous, 1951; Soldo and Jardas, 2000a). A few years later, on October 2, 1954, an attack on a boat occurred off Pag. This shark was estimated to measure 550 cm and to weigh 1500 kg (Anonymous, 1954; Soldo and Jardas, 2000a). The next year, on December 10, a 460 cm female was caught off Senj, Sveti Juraj. This specimen is preserved as a taxidermied skin-mount at the Zemaljski Muzej Bosne i Hercegovine in Sarajevo, Bosnia and Herzegovina (without cat. no.) (De Maddalena, 2006b, 2007; Đurović and Obratil, 1984, 1989; Drazen Kotrošan, pers. comm.). In 1956, a 400 cm shark was reported off Krk, but it is unclear if it was captured. The record was reported by Soldo and Jardas (2002) based on personal communication they received from Kovačič.

The 1960s started with an attack on September 24, 1961, that occurred off Opatija. In the early afternoon, a 19-year-old boy, Sabit Plana, was swimming 70 m offshore where he was attacked by a large shark. A boat was deployed to rescue him. But by the time it reached him, he had lost his left hand, had both legs bitten, and was already deceased (Anonymous, 1961; Baldridge, 1973; Giudici and Fino, 1989; GSAF). On August 16, 1966, another fatal attack occurred in an unknown location off the Dalmatian coast (Angelo Mojetta, pers. comm.). In 1968, a juvenile white shark was caught in a gill net north of Rava in approximately 40 m-deep water. The specimen was estimated to be about 200 cm TOT by author De Maddalena on the basis of the photographic evidence that still exists (Robert Lončarić, pers. comm.; Lončarić, 2009). In 1970 an attack occurred on a diver off Novigrad. Mr. Jurincic, the diver, was spearfishing (Angelo Mojetta, pers. comm.). A year later, in 1971, a white shark attack occurred off Ika. The victim, Stanislaw Klepka, was swimming when the shark attacked him, severing his leg and causing his death by exsanguination (Gilioli, 1989). On August 17, 1972, a 600 cm shark was reported off Kornati, but it is unclear if it was captured (Morovic, 1973; Soldo and Jardas, 2002). Two years later, on August 10, 1974, an attack occurred off Omiš. At about 3:00 p.m., Rolf Schneider was attacked by an estimated 500 cm shark that killed him by severing a foot and also cutting the femoral artery. This case is probably the same case reported by Jolic (1988) and by Soldo and Jardas (2002) with a different location (Lokva Rogoznica) (Marco Zuffa, pers. comm.).

Almost two decades transpired before another white shark was reported from Croatia. In August 1993, a 500 cm-long female shark was caught in a net off Sibenik. This was reported by Fergusson (1996) based on a personal communication he received from G. Notarbartolo di Sciara. Also in August 1993, a large shark was sighted several times by fishermen off Losinj. Fergusson (1996) reported this record also based on a personal communication he received from G. Notarbartolo di Sciara. This case is surely the same case reported by Mojetta et al. (1997) with a different location (Cres). In the spring of 1994, a shark was caught off Brac and brought to the fish market in Split. The jaws were examined by Gianfranco Della Rovere, in the summer of 2004, who estimated the jaw diameter at about 30 cm. Della Rovere reported this record on the basis of a personal communication he received from the fisherman who captured the shark (Gianfranco Della Rovere, pers. comm.). On August 2, 1998, a shark was sighted off Mljet (De Sabata et al., 1999). Another shark was sighted a month later, on September 4, 1998, off Dubrovnik (De Sabata et al., 1999).

In the spring of 2000, an estimated 600 cm-long shark was sighted by a fisherman off Porec. This record was reported by Daniel Abed-Navandi on the basis of a personal communication he received from Mr. Dario, the commander of the ship *Burin* (Daniel Abed-Navandi, pers. comm.). On June 24, 2003, a large female white shark was caught 15 miles southwest of Jabuka. The shark was entangled in a tuna net set by Pave Arkovic from the fishing boat *Ponos*. In order to prevent damage to the net, the fishermen attempted to chase the shark away, but without success. The entangled shark was hauled aboard the boat, where it struggled for four hours before it finally died. The fishermen kept some of the teeth and the carcass was then thrown back into the sea. It was reported measuring 570 cm and estimated to weigh about 2500 kg. Photographic documentation of this shark exists (Lovrenc Lipej, pers. comm.; Soldo and Dulcic, 2005). Recently, on October 6, 2008, an attack occurred in Mala Smokova Bay, off Vis. In the morning, 43-year-old Slovenian diver Damjan Pesek shot a greater amberjack *Seriola dumerili* and attached the bleeding fish to his waist. Pesek was at the surface 10 m from the shore in water that was only 10 m deep when an estimated 450–500 cm great white shark attacked, biting him on the lower part of the left leg. Pesek grabbed the shark by the gill slits with one hand and hit its head with the other hand. The shark released him and he was able to get back to the dive boat. The wound was serious, but the bleeding was stopped aboard the boat and the victim was air lifted to the hospital in Split by helicopter. Two white shark tooth fragments where found in the wound and measured less than 1 cm-long, allowing researcher Alen Soldo to confirm the species identification (Branko Dragicevic, pers. comm.; Jakov Dulcic, pers. comm.).

Bosnia and Herzegovina

The great white shark data bank has no records from Bosnia and Herzegovina.

Montenegro

Only three records from Montenegro are recorded in the data bank. In June 1926 a 300 cm shark was caught off Herceg Novi. Its stomach contained a woman's shoes and clothes (Hirtz, 1932; Morovic, 1950; Soldo and Jardas, 2002). On August 5, 2011, a 217 cm shark was accidentally caught in a fishing net about 30 km off Bar. Photographic documentation of this speciment exists (http://sharkyear.com). In August 1955 a fatal attack occurred off Budva (Radovanovic, 1965; Soldo and Jardas, 2002).

Albania

The great white shark data bank has no records from Albania.

Greece

The first record in the data bank from Greece occurred some years before 1878, when an estimated 500 cm shark was caught off Syros, Cyclades Islands. The shark ventured close inshore to eat some leather that was placed in the water near the shore close to a tannery. The source (Heldreich, 1878) erroneously identified it as a porbeagle *Lamna nasus*. This specimen was said to be preserved at a museum in Athens. Three sharks were reported before 1943. The first shark was recorded off Alexandroupolis, but it is unclear if it was captured (Fergusson, 1996; Konsuloff and Drenski, 1943). The second was a 230 cm shark that was recorded off Kavallah, but it is also unclear if it was captured (Fergusson, 1996; Konsuloff and Drenski, 1943). The third shark measured 180 cm in length and was recorded off Thasos, and again it is unclear if it was captured (Fergusson, 1996; Konsuloff and Drenski, 1943).

On September 22, 1948, an attack occurred off Keratsini, Peiraus Port, in the Saronic Gulf. At 4:00 p.m., a 17-year-old boy, Dimitris Th. Parasakis, was jumping from the shore into the water and swimming within 5 m of the shoreline when an estimated 640 cm white shark fatally attacked him. The shark attacked twice, severing the right arm and hand on the first attack and biting the boy from the head to the lumbar region on the second attack. This incident was reported by other sources with the wrong name of the victim (Dedemetri Valasakis) (Manolis Bardanis, pers. comm.; Coppleson, 1958). Another attack occurred on August 17, 1951, off Mon Repo in Kerkira. At noon, a 16-year-old girl, Vanda Pierri, the daughter of the director of the national bank in Kerkira, was swimming with an 18-year-old boy, George Athanasenas. The depth of water was 6–7 m. They were both attacked by a great white shark that fatally cut Pierri in pieces and injured Athanasenas. Despite the fact that the incident shocked the people of Kerkira, the Athens newspapers dedicated only a few lines to the attack in the back pages of the paper (Manolis Bardanis, pers. comm.). Another attack occurred in 1956, off Kerkira (Corfu). A 15-year-old girl, Ms. Margoulis, was fatally attacked by a white shark while swimming off a yacht (MEDSAF).

The next report of a white shark in Greek waters was on June 4, 1962, at 8:30 a.m., when a shark was caught off Aliki-Kato Axaia. It was reported measuring 480 cm and weighing 2000 kg. This specimen was photographed and the documentation still exists (Manolis Bardanis, pers. comm.). On June 1, 1963, an attack occurred just 3 m off Pithos or Micra Rock, near Trikerion Island, Pagasitikos Gulf, which is near Volos. A few minutes after 4:30 p.m. a 42-year-old Austrian woman, Helga Pogl, was swimming in waters 80 cm deep when she was fatally attacked by a white shark estimated to measure 300 cm in length, that took her away into deeper water. Her friend Wilgen witnessed the attack. Pogl's body was never found (Manolis Bardanis, pers. comm.). Another fatal attack occurred in 1969 at an unknown location in Greek waters. A 15-year-old had just jumped from the shore into the water and was swimming when he was fatally attacked by a great white shark (Manolis Bardanis, pers. comm.). On September 15, 1972, a male shark was caught 600 m off Makryalos, Thermaïkos Gulf. It was reported measuring 460 cm and weighing 1300 kg (Economidis and Bauchot, 1976).

Over a decade later, around 1985, a large white shark was caught off Paliouri, Halkidiki (Manolis Bardanis, pers. comm.). It was estimated to be 601–618 cm TOT by author De Maddalena on the basis of the photographic evidence that still exists. On February 10, 1991, an estimated 600 cm white shark was caught on a longline set 7 miles off Kerkira (Corfu). Its stomach contained an intact loggerhead sea turtle *Caretta caretta* having a 60 cm-diameter carapace. The set of jaws was bought by a private collector and is presumably still preserved in his private collection. The rest of the shark was sold for human consumption to a restaurant. Photographic documentation of the jaws exists (Marco Zuffa, pers. comm.). In August 1999, an estimated 400–500 cm shark was encountered 2 miles off Othoni (Fano). The shark was seen swimming on a shoal with a bottom depth of 70–80 m. The weather was good and the sea was flat. The shark approached an inflatable boat and carefully observed the occupants in the boat before they decided to leave (L. Leone De Castris, pers. comm.).

Turkey

The earliest record of a white shark in the data bank from Turkey was in February 1881, when a shark was stranded near Beylerbeyi, in the Bosporus. It was landed in Serraglio Point, Istanbul, and was reported measuring 391 cm and having a girth of 335 cm (Fergusson, 1996). On November 17, 1881, a female shark was caught in the Bosporus. It was reported measuring 470 cm and weighing 1500 kg (Fergusson, 1996).

In 1916, a large shark estimated at 700 cm was caught in a tuna trap off Salistra, in the Sea of Marmara. The shark entered the tuna trap near Fenerbahçe Harbor and became trapped in the nets, where it was killed with three rifle shots to the head. The carcass was gutted, cut up and sold at the fish trap because it was too large to be transported to the fish market (Devedjian, 1945; Kabasakal, 2003). In May of 1920, a shark was caught off Sedef Adasi, in the Sea of Marmara. It was reported measuring 465 cm and weighing about 1200 kg. It was caught with a line intended for swordfish and was brought to the Istanbul fish market, where it was put on display for a long time. The stomach contained an estimated 200 kg bluefin tuna *Thunnus thynnus*, remains of a swordfish *Xiphias gladius*, some Atlantic bonito *Sarda sarda* and a small stone (Devedjian, 1945; Kabasakal, 2003).

Before 1926, a shark was caught with hand line intended for tuna in the Sea of Marmara. The shark was brought to the Istanbul fish market and was reported to measure about 400 cm with a girth of 135 cm. Its stomach contained eight large Atlantic bonito *Sarda sarda* (Devedjian, 1945; Kabasakal, 2003). In the same year on February 20, a shark was caught off Büyükada, in the Sea of Marmara. It was reported to measure 450 cm and weighed over 1500 kg. Like some of the other sharks, this one was caught by a fisherman with hand line that was intended for swordfish. Kabasakal (2003) reported this record on the basis of a personal communication he received from Agop Savul.

In 1930 an attack occurred in the Bosporus. The shark attacked a boat occupied by two persons of English nationality and hit the hull so hard that it momentarily got its head stuck (Anonymous, 1930; Xuereb, 1998). According to Xuereb (1998), it may have been a great white shark. However, there is some doubt over the exact identity of the species. On March 30, 1954, a shark was caught off Tuzla in the Sea of Marmara

2—Records of Great White Sharks from the Mediterranean Sea

by a fisherman with a tuna hand line. This shark was brought to the Istanbul fish market, where it was displayed for quite some time. It was reported measuring 450 cm and weighing 1500 kg. Kabasakal (2003) reported this record based on a personal communication he received from Agop Savul. Two years later on April 15, 1956, a large female shark was caught near Prince Islands, in the Sea of Marmara. Caught by fisherman Aziz Ünlü with hand line for tuna, the shark was hooked in the early morning and harpooned after a seven-hour fight. It was reported measuring 618 cm and weighing about 3000 kg. Kabasakal (2003) reported this record on the basis of a personal communication he received from Agop Savul.

In February 1962, a female shark was caught in the Bosporus. It was reported measuring over 500 cm and weighing 3750 kg, but according to Fergusson (1996), the weight may be wrong. Fergusson (1996) reported this record based on a personal communication he received from G. Wood. It may be the same shark portrayed in a photograph, published in the July 1967 issue of *Mondo Sommerso* magazine, reported

Top: Two young boys pose with this 450 cm white shark was caught on March 30, 1954, off Tuzla, Turkey (photograph from Agop Savul's archive, courtesy Hakan Kabasakal). *Bottom:* This estimated about 600 cm female white shark was caught on December 28, 1965, off Maiden's Tower, Turkey (courtesy Hakan Kabasakal).

to be caught by fisherman Fahri Gur, but with a different length (450 cm). Three years later, on December 28, 1965, a female shark was caught off Dolmabahçe, in the Bosporus. It was caught by three fishermen with a hand line fishing for tuna and after a long fight was harpooned. The shark was landed on the Dolmabahçe coast and was reported measuring 500 cm and weighing about 4000 kg (that seems huge for a 500 cm shark). Kabasakal (2003) reported this record on the basis of a personal communication he received from Agop Savul. On the same day, another female shark was caught in the Bosporus but off Maiden's Tower. It was caught by Hüseyin Şalvarli with a tuna hand line. The shark towed the boat for a long time, but in the end Şalvarli succeeded in gaffing the shark with the anchor from his boat. This was a large shark, reported measuring 700 cm and weighing about 3000 kg, even though analysis of the photograph by author De Maddalena seems to suggest that it did not measure over 600 cm TOT. Kabasakal (2003) reported this record on the basis of a personal communication he received from Agop Savul.

These two white sharks estimated at 400 cm each (the one on the left is on its back with its white belly up and is barely visible) were caught on January 13, 1966, off Kabataş, Turkey (photograph from Agop Savul's archive, courtesy Hakan Kabasakal).

On January 13, 1966, two sharks were caught by Hakki Baba and Ali Yavur off Kabataş, in the Bosporus. After four and half hours of fight, the sharks were harpooned near Kabataş. Both were similar in size, reported to measure about 400 cm and to weigh about 2000 kg. Kabasakal (2003) reported this record based on a personal communication he received from Agop Savul. In 1967, a shark was caught off Büyükada, in the Sea of Marmara, by fishermen with tuna hand line. However, the end of the line had a 30–40 cm-long-shank shark hook. The shark was hooked off the southern coast of Büyükada and after a 13-hour battle was landed near Salacak pier. Photographic documentation of this shark exists (Kabasakal, 2008). In late March 1968, an estimated 551 cm female great white shark was caught by a tuna hand-liner in the Bosporus Strait. Photographic documentation of this shark exists (Habasakal, 2011).

Before 1974, an estimated 2000 kg shark was caught near Prince Islands in the Sea of Marmara by fishermen using a hand line for tuna (Güney, 1974; Kabasakal, 2003). A decade later, in May 1985, an estimated 500 cm-long shark was encountered by off Kapidağ in the Sea of Marmara. The shark circled a fishing boat for a few minutes,

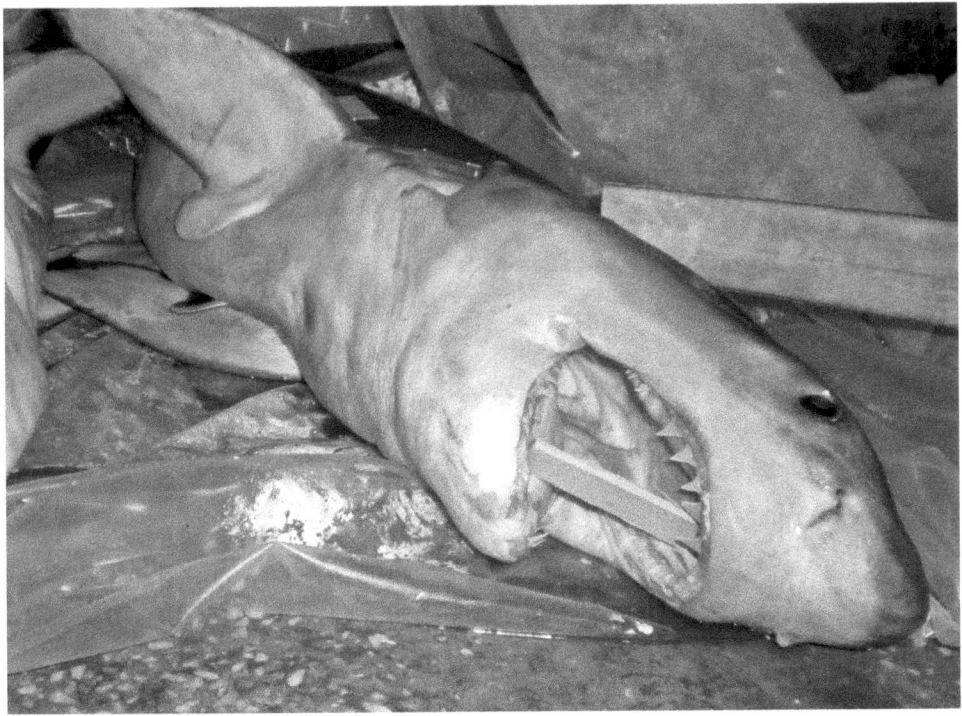

This approximately 130 cm-long newborn white shark was caught in early July 2008 in the Edremit Bay, Turkey (photograph by Hakan Kabasakal).

then it left (Kabasakal, 2003). On March 18, 1991, a shark was caught in 120 m deep water, 6 miles off Foça, in the Aegean Sea. It was reported to measure about 550 cm and to weigh about 3000 kg. The stomach contained a tuna estimated to be 100 cm-long. The shark was brought to the Istanbul fish market for sale at an auction. It was photographed and the documentation still exists (Anonymous, 1991a; CLOMFOT website; Kabasakal, 2008). Five years later, in March of 1996, a 550 cm female was caught in a purse seine near Gökçeada Island in the Aegean Sea (Kabasakal and Kabasakal, 2004). Two years later, in April of 1998, an estimated 450 cm-long shark was sighted by a fisherman offshore to the west of Gökçeada Island (Kabasakal and Kabasakal, 2004). In May 1999, a shark estimated at 500 cm was encountered by a diver in the Gulf of Saros, in the Aegean Sea (Kabasakal and Kabasakal, 2004).

Most recently, on July 1, 2008, a 125.5 cm TL newborn white shark was caught in a gill net one km off Altınoluk in the north Aegean Sea. Three days later, on July 4, 2008, another small shark measuring 145 cm TL was caught on a bottom longline in the same place, one km off Altınoluk. The fishermen tried to keep both great white sharks alive for display in a 25-ton marine aquarium. However, both sharks were short-lived, the first dying only 12 hours after capture and the second only 27 hours after capture. After the death of both specimens, they were preserved with ice and transported to the İstanbul fish market for delivery to the Ichthyological Research Society. Morphometric measurements of the specimens were reported in Kabasakal and Gedikoğlu (2008). Jaws and caudal fins of these specimens are preserved at the Ichthyological Research Society in Istanbul. Both specimens had clear (as in transparent or noticeable) umbilical scars on their throats, which indicated they were newborn. The stomach of the smaller shark included many embryonic teeth. The stomach of the 145 cm specimen contained a few teeth and remains of a bony fish (probably the bait) (Kabasakal and Gedikoğlu, 2008).

Syria

The great white shark data bank has no records from Syria.

Cyprus

The great white shark data bank has only one record from Cyprus. In 1993, a shark estimated at over 500 cm, possibly a female, was encountered by divers on several occasions (Fisher et al., 1987).

Lebanon

The great white shark data bank has no records from Lebanon.

Israel

The great white shark data bank has only one record from waters off modern-day Israel. Before 1971, a 200 cm shark was caught by hook and line off Akko. The jaws were preserved (Ben-Tuvia, 1971).

Egypt

The great white shark data bank has only one record from Egypt. In the summer of 1934, a female shark was caught off Al Iskandariyya (Alexandria) and landed in Agamy Beach. It was reported measuring 425 cm and weighing 2500 kg. The shark was a pregnant female carrying nine embryos that were reported measuring 61 cm and weighing 49 kg (108 lbs.) (Norman and Fraser, 1937). The weight is obviously wrong.

Tricas and McCosker (1984) hypothesized that the original source made a mistake and that 49 kg was the total weight of the nine embryos, so the average weight of each embryo was 5.4 kg. Another possibility is that the correct weight for each embryo was 4.9 kg, which was written as 49 kg by mistake. Photographic documentation of this specimen exists.

Libya

On June 12, 2002, a 50 m-diameter tuna pen belonging to a European tuna farm company was being towed from Libya to Spain. In the morning, the tuna cage was located at 33º 50′N, 13º 50′E, 55 miles off Tripoli, where the towing boat had stopped to check the tuna cage. During the previous five days, bad weather had prevented the tuna farm staff from checking the cage, but on that day the weather was sunny and calm. Biologist Txema Galaz, accompanied by some other divers, prepared to check the cage, which contained 60 tons of bluefin tuna *Thunnus thynnus*. When Galaz dove into the cage and conducted the first check, he did not notice anything unusual, and he ordered some divers to go to the bottom of the net to remove some dead tuna. It was about 10:00 when the divers descended to the bottom of the pen, where they saw a large white shark breaking through the netting at the bottom. It took only five seconds for the shark to enter the tuna cage. The white shark had likely been following the tuna cage as it was being towed for some time before it broke through the net when the towing was suspended for inspection.

As soon as the shark entered the cage, the tuna became suddenly excited and the divers left the cage immediately. Galaz and the rest of the tuna farm staff watched the shark for about two and half hours. The shark started to swim in a circle inside the cage, initially swimming about 20 m deep, then ascending closer to the surface. About five minutes after it broke into the pen, the shark attacked the net, biting it for only a few seconds. Then, it continued to swim, showing no aggression toward the net and no desire to break through. The large predator appeared relaxed and simply continued to swim about 1 to 2 m below the surface near the side of the cage, coming to the surface a few times. Its total length was estimated at about 500 cm. Concerning the gender, although it was observed for a long time, claspers were not visible. If the gender was male, the claspers would be very long and well developed for a male of this size; therefore it can be speculated that it was a female.

A couple of underwater photographs of the white shark were shot by submerging a camera into the water from the surface. Some additional topside photos of the shark swimming slowly in the cage with its dorsal fin protruding above the surface were also shot. Additionally, a low-quality video was shot. Surprisingly, the fear reaction of the tuna to the presence of the shark lasted only for a short time. The tuna did not seem to be nervous of the presence of the large predator so close to them, but they swam as far away as possible from it. No tuna escaped through the relatively small hole made by the shark when it entered the cage from the bottom (the tuna simply did not swim close to the net, which is their usual behavior). The shark was not observed preying on or attacking any of the tuna, but we cannot be sure whether it fed on one or more tuna during its entire time inside the cage. Possibly the shark was more concerned about being trapped in the cage and paid little attention to its potential prey. However, for all the time it was observed, it made no attempt to break the net, even though this was surely not a big obstacle for it, considering how easily it broke the net when it entered the pen. The tuna farm staff decided to leave the cage alone and ordered the towing boat to continue its travel. When the boat restarted the tow, the shark seemed uncomfortable in the cage. Two days later the shark was gone from the tuna cage. It is uncertain when the shark escaped from the cage, but it made a hole in the net on the vertical wall of the cage (Galaz and De Maddalena, 2004).

An estimated 500 cm female white shark swims in a tuna cage on June 12, 2002, off Tripoli, Libya (photograph by Lorenzo Millan).

Malta

The earliest records in the great white shark data bank from Malta are from the 1800s. On April 25, 1890, a fatal attack on two persons occurred off the Munxar reefs. At 7:00 a.m., 11 km off Marsaskala Bay, an animal hit the boat, capsizing it and causing fishermen Salvatore Bugeja, Agostino Bugeja, Carmelo Delia and Carmelo Abela to fall into the sea. Delia and Abela were rescued by a boat that was passing nearby, but Salvatore and Agostino Bugeja were never found. There is some doubt over the identity of the species. Gulia (1890) hypothesized that it was a great white shark. However, a report said that immediately before the animal attacked, it produced a loud cry. Therefore, a cetacean cannot be excluded as the possible attacker. (The first species that comes to mind is a killer whale *Orcinus orca*. However, its occurrence in the Mediterranean Sea is exceptionally rare and attacks on boats by this species are also extremely rare.) In addition, there is a testimony of this case by an ex-voto preserved in the Zabbar Sanctuary Museum, which was reproduced in Xuereb (1998). Three days after the attack, on April 28, 1890, a shark was sighted that may have been the animal responsible for the attack. It was likely attracted by some carcasses floating in a bay close to the location of the sighting. It was reported that a shark, possibly the same shark, had been encountered a couple months earlier, on February 20, 1890, by three fishermen in an unknown location 4 km off Malta (Gulia, 1890; Xuereb, 1998). On May 14, 1898, a 1429 kg great white shark was caught in a tuna trap off Mellieha. The liver alone weighed 159 kg. It was brought to Valletta and displayed to the public for a viewing fee. Afterwards, it was sold for human consumption. Even though it was likely a white shark, there is some doubt over the exact identity of the species (Anonymous, 1898).

An attack occurred on March 7, 1907, off Marsaskala. An estimated 600 cm shark attacked a boat, causing two fishermen to fall into the sea where it then fatally attacked them (Mojetta et al., 1997). In September 1914, a juvenile shark was caught off Wied iz-Zurrieq. It was sold at the fish market in Valletta (Xuereb, 1998). On July 20, 1956, an attack occurred in St. Thomas Bay. The weather was good, with a light breeze, and the water temperature was about 30°C. At 4:30 p.m., Jack Smedley and Tony Grech were swimming 1.5 m from each other about 100 m off Munxar when Smedley was attacked by an estimated 500 cm-long great white shark. Smedley disappeared under the surface. He emerged briefly and then disappeared forever. Grech reached the shore unharmed. Alfred Xuereb was on the shore and saw the attack from above the bay. A nearby fisherman said that he sighted the shark swimming towards Munxar prior to the attack. Remains of the victim were never fround (Alfred Xuereb, pers. comm.). Gianturco (1978) reported it with a wrong date (July 21) and a wrong location (Valletta). Mojetta et al. (1997) reported this case with a wrong date (July 18). Another attack occurred in August of 1956 in the Congreve Channel, between Filfla and Wied iz-Zurrieq. A fishing vessel off Filfla was hauling its net aboard when a white shark emerged from the depths and rammed the boat broadside with its head, causing the boat to capsize. Two fishermen, Nazzareno Zammit and Emmanuel, fell into the sea where Emmanuel was fatally attacked by the shark (Marco Zuffa, pers. comm.). According to other sources, the shark did not cause any injury to the men, but Emmanuel later died of shock at the hospital. The GSAF reported this record on the basis of a personal communication received from Alex Buttigieg.

On an unknown date between 1960 and 1969, a white shark bit the drop keel of a yacht anchored in the Congreve Channel near Filfla. The GSAF reported this record on the basis of a personal communication received from Alex Buttigieg. In May of 1964, a male shark was caught with a line off Filfla. The animal was reported by Fergusson (1996) to weigh approximately 2000 kg and to measure approximately 500 cm. However, it was reported to measure 580 cm by Alfred Xuereb (pers. comm.). Photographic documentation of this shark exists. Fergusson (1996) reported the record based on a personal communication he received from John Abela. Nine years later, in March of 1973, a shark was caught in a net fished near the ocean bottom off Filfla. The shark was reported measuring 530–550 cm and weighing 1920 kg. Two different versions of this capture were reported. According to Mizzi (1994), the shark was a female and was caught on March 30, 1973, near Filfla by brothers Emanuele and Andrea D'Amato. It was landed in Wied-Iz-Zurrieq and was reported measuring 549 cm and weighing 2000 kg. The stomach contained a 61 cm sea turtle, a stone weighing approximately 7 kg, a 274 cm-long metallic cable and 31 rusted hooks measuring 15 cm each. According to Alfred Xuereb (pers. comm.), the shark was a male and was caught on March 3, 1973, by Emanuel D'Amato, Matty Cachia and Alfredo Cutajar. It was brought to the fish market in Valletta. Photographic documentation of this shark exists (Abela, 1989; Mizzi, 1994; Fergusson, 1996; Mojetta et al., 1997; Alfred Xuereb, pers. comm.). On March 4, 1979, a 400 cm-long male shark was caught by a fisherman nicknamed Popeye off Filfla. Photographic documentation of this shark exists (Alfred Xuereb, pers. comm.). Fergusson (1996) reported this record with a different date (May 1979) on the basis of a personal communication he received from John Abela and stated that it was a female.

The first encounter in the 1980s was in 1982, when an estimated 500 cm shark was sighted off Gozo (Mojetta et al., 1997). On April 17, 1987, a large female was caught off Filfla. The capture occurred in the morning and was made by fishermen Alfredo Cutajar and Vince D'Amato three miles off Blue Grotto, south of Filfla. The shark was initially hooked with a fixed surface longline belonging to Mr. D'Amato, but the line snapped and became entangled with a line belonging to Mr. Cutajar, who eventually landed the shark. The shark was taken to Wied-Iz-Zurrieq, but it was too heavy for the harbor winch. So it was landed in Marsaxlokk. Its stomach contents included a whole 220 cm-long blue shark *Prionace glauca*, a 250 cm dolphin of unidentified species that was consumed in two (or three, according to some other sources) parts, a loggerhead sea turtle *Caretta caretta* having a 60 cm-diameter carapace, and a plastic bag containing garbage. Other sources reported the sizes of the contents as follows: blue shark 183 cm, dolphin 244 cm, marine turtle 61 cm. That evening, the shark was stored in a large garage in Zurrieq. The next day, April 18, 1987, it was taken to the Valletta fish market. The jaws were preserved by John Abela and are exhibited in the Museum of History and Culture on Gozo Island. The case was reported and discussed by several authors (Abela, 1989; De Maddalena, 1999, 2001; Ellis and McCosker, 1991; Fergusson, 1996, 1998b; Fergusson et al., 2000; Mollet et al., 1996). Unfortunately, no fisheries officer measured this large shark (Michael Darmanin, pers. comm.). Abela (1989) reported measuring the total length of the shark accurately in a straight line (as TOT, according to Fergusson, 1998) at 714 cm. He also specified a weight of 2730 kg. That size is slightly smaller than that reported in a letter written by the shark's capturer, Alfredo Cutajar, in May 1991 to Mr. Giuliano Chiocca. Cutajar stated that it was 723 cm-long and weighed 2880 kg.

A few years later, in the mid–1990s, Alfredo Cutajar and John Abela were interviewed for the Jeremy Taylor documentary *Jaws in the Med*. Mr. Cutajar reported a length of 23 feet (701 cm) for this specimen, while Abela confirmed that he had accurately measured the shark twice, as it lay flat on the floor, at 23 feet 5 inches (714 cm) long. Recently, in 2001, Mr. Cutajar stated he estimated the shark as being 23 feet (701 cm) long based on the length of the pickup truck on which it was placed with its caudal fin hanging out (this statement clearly indicates that he did not measure the length of the shark). Moreover, he declared that the approximate weight was 3 tons, which was estimated on the basis of the sum of the weights of the individual parts of the butchered shark that were sold at the fishmarket (Michael Darmanin, pers. comm). Mollet et al. (1996) wrote that on the basis of discordant testimonies by John Abela, the shark's length was only estimated. Furthermore, Fergusson (1998b) stated that the length was in the 520–550 cm TL range, a conclusion he reached after examining the photograph of the shark taken by photographer John Gullaumier and published by the Maltese newspaper *In-Nazzjon-Taghna* (Anonymous, 1987). This conclusion was also confirmed by a forensic investigator who analyzed the picture at the request of the BBC Natural History Unit.

Spanish shark researchers Joan Barrull, right, and Isabel Mate, of the Museu de Zoologia in Barcelona, Spain, with the set of jaws from an estimated 668–681 cm female white shark caught on April 17, 1987, near Filfla, Malta (photograph by Joan Barrull and Isabel Mate).

A few years later, Fergusson et al. (2000) again confirmed this estimate, stating that the shark was no longer than 550 cm TL. This length agrees with the size reported by Anonymous (1987), who gave the shark's weight as 1.5 tons and its length as "about 18 feet" (549 cm). According to Fergusson (1998), Abela was later interviewed by the BBC, and he admitted that he may have made a mistake in taking the measurement. However, in 2001, Abela reconfirmed the 714 cm total length but added that he had taken the shark's length measurement with a rope, which he eventually measured later (Michael Darmanin, pers. comm.). The length of "about 18 feet" reported by Anonymous (1987) was not a measurement but merely an estimate (Alex Buttigieg, pers. comm.); similarly, the same may be true of the 1.5 ton weight reported by the same source (Anonymous, pers. comm.).

In the end, it was not possible to find anyone else who may have measured the shark, so it seems that John Abela was probably the only person to have measured it. Three eyewitness testimonies were collected from those who had seen the shark. The eyewitnesses were Mr. Alex Buttigieg, a person who asked to remain anonymous, and the fisheries officer Mr. Grezzju Grech. None of them measured the shark. The anonymous witness went to the Valletta fish market early in the morning with a camera and measuring tape only minutes before the shark was cut up but was denied permission to take photos or measurements. Alex Buttigieg and the anonymous witness declared to have estimated the length to be significantly less than the 714 cm TOT reported by John Abela but larger than the 520–550 cm TL estimated by Fergusson (1998) and Fergusson et al. (2000). The anonymous eyewitness estimated the length at approximately "18.5–19.5 feet" (564–594 cm), or "close to 20 feet" (610 cm). This estimate was made at a distance of about 12 feet from the shark, which had been placed on the floor of the Valletta fish market. Alex Buttigieg estimated it to be "less than 20 feet" (610 cm) from a distance of about 20 m. Mr. Grezzju Grech, who was present when the shark was landed, affirmed that in his opinion it was possible that the shark was 7 m long (Michael Darmanin, pers. comm.).

There were several requests for more information about this case made recently by author De Maddalena to John Abela that remained unanswered. However, there were some replies indirectly from John Abela via Mr. Michael Darmanin, senior fisheries officer at Malta Centre for Fisheries Sciences (Department of Fisheries and Aquaculture, Fort San Lucjan, Marsaxlokk). This recent information, together with the testimony from Mr. Buttigieg and the anonymous witness, helped to reconstruct the entire story. Ultimately, this shark was estimated to be 668–681 cm TOT and 647–660 cm TLn by De Maddalena et al. (2001) on the basis of the photographic evidence. De Maddalena et al. (2001) concluded that there was insufficient evidence to refute the 714 cm TOT indicated by John Abela. If Abela made an error measuring the shark, which is possible considering the way he measured the shark and in view of the results obtained by this study, the true length TOT of the shark was not significantly shorter (33–46 cm based on the estimates) than the length he reported. According to the estimates by

De Maddalena et al. (2001) and the observations made by the five eyewitnesses (Abela, Cutajar, Grech, Buttigieg, Anonymous), the 520–550 cm TL indicated by Fergusson (1998) and Fergusson et al. (2000) is unacceptably short.

Around April 1987, an estimated 600 cm male shark was encountered off Malta. A fisherman armed with a steel leader and a 25 cm-wide hook attempted to catch the shark. On August 18, 1993, a male shark in excess of 500 cm-long approached a boat off Malta. Fergusson (1996) reported this record on the basis of a personal communication he received by I. Maxwell. The last record from Malta is on an unknown date, when a great white shark was caught off Malta. Three of its teeth are preserved at the Zoologisk Museum in Copenhagen, Denmark (without cat. no.) (De Maddalena, 2006b, 2007; Jørgen Nielsen, pers. comm.).

Tunisia

From 1953 to 1976, many sharks were recorded in the Golfe de Gabès off Kelibia and La Galite (Capapé et al., 1976; Fergusson, 1996). Around May 20, 1953, a shark was caught in a tuna trap off Sidi Daoud. It was reported measuring about 600 cm and weighing 1500 kg (Postel, 1958). A year later, on May 20, 1954, two sharks were caught in a tuna trap off Sidi Daoud (Postel, 1958). On May 16, 1956, a female shark was also caught in a tuna trap off Sidi Daoud. It was reported measuring 520 cm and weighing 1800 kg. The stomach contents included skeletons of two common dolphins, *Delphinus delphis* (an adult and probably an embryo), parts of a green sea turtle, *Chelonia mydas*, and a small mako shark, *Isurus* sp. (most likely a shortfin mako *Isurus oxyrinchus*), that measured a little over one meter. Postel (1958) reported some morphometric measurements, and photographic documentation of this shark exists. On May 22, 1956, a male shark was caught in a tuna trap off Sidi Daoud. It was reported measuring 410 cm and weighing 1300 kg. The source, Postel (1958), reported some morphometric measurements. Over a decade later, around 1970, some sharks were caught near Zembra Island and off Bizerte (Capapé et al., 1976; Fergusson, 1996). Between 1970 and 1980 a 523 cm male was caught off Ras Fartas and was brought to the fish market in Tunis (Bradaï et al., 2002). On March 21, 1972, two 400 cm sharks were caught off Zarzis (Quignard and Capapé, 1972). A month later, on April 21, a small shark measuring 185 cm was caught off Zarzis (Quignard and Capapé, 1972).

In May 1979, a 220 cm female shark was reported northwest of Cap Bon, but it is unclear if it was captured. Fergusson (1996) reported this record based on a personal communication he received from Christian Capapé. A month later, in June, an estimated 520 cm-long female shark was caught northwest of Cap Bon. The stomach contained a dolphin. Fergusson (1996) reported this record on the basis of a personal communication he received from J. Zaouali. In August 1981, a small male shark measuring 185 cm was caught northwest of Cap Bon. Fergusson (1996) reported this record

This 520 cm female white shark was caught on May 16, 1956, off Sidi Daoud, Tunisia (photograph from Postel, 1958, courtesy Muséum National d'Historie Naturelle de Paris).

This 520 cm white shark was caught on April 27, 2001, off Sidi Daoud, Tunisia (photograph by Walid Maamouri).

This 587 cm pregnant female white shark, carrying 4 embryos ranging from 132 to 135 cm in length, was caught on February 26, 2004, in the Gulf of Gabès, Tunisia (photograph by Mohamed Nejmeddine Bradaï).

on the basis of a personal communication he received from Christian Capapé. In June 1983 another small shark was caught northwest of Cap Bon. This female shark measured 175 cm in length. Fergusson (1996) reported this record based on a personal communication he received from Christian Capapé. In April 1985, another small female shark measuring 190 cm was caught northwest of Cap Bon. Fergusson (1996) reported this record on the basis of a personal communication he received from Christian Capapé. In July 1988 two males were caught in the Golfe de Gabès. These were small sharks, one measuring 220 cm and the other measuring 165 cm (Bradaï et al., 2002). On February 4, 1989, a female shark was caught off Ghar El Melh. The shark was following a fishing boat, so they captured it (Anonymous, 1989a; Giudici and Fino, 1989). Two different lengths were reported for this shark: 670 cm and 570 cm. The weight was reported at 2032 kg. Fergusson (1996) reported this case with two wrong dates (April 1 and January 4, 1989).

In 1992, perhaps in September, a great white shark estimated at over 500 cm was caught northwest of Cap Bon. This shark was a pregnant female carrying two full-term embryos. Neither the female nor the embryos were examined or measured by researchers. Fergusson (1996) reported this record on the basis of a personal communication he received from Christian Capapé. On December 14, 1992, a 435 cm male shark was caught by a tuna boat using a purse seine in the Golfe de Gabès (Bradaï et al., 2002). A few months later, on February 18, 1993, a similarly sized male shark measuring 425

This young female white shark was caught on April, 20, 2006, off Aras Dizra, Tunisia (photograph by Samuel P. Iglésias/Muséum national d'Histoire naturelle de Paris).

cm was caught by a tuna boat using a purse seine in the Golfe de Gabès (Bradaï et al., 2002). In August 1993, a shark was encountered on the Skerki Bank. It was seen swimming at the surface in the evening by shrimp fishermen. The shark hit the small boat violently, so the fishermen transferred to a larger boat that was in close proximity. This shark was large, estimated to measure over 600 cm. Photographic documentation of the shark exists (Fergusson, 1998c). On February 9, 1997, a 425 cm female shark was caught by a tuna boat using a purse seine in the Golfe de Gabès (Bradaï et al., 2002). On February 26, 1997, a 179 cm male was caught by a trawler in the Golfe de Gabès (Bradaï et al., 2002) and on December 25, 1997, another small male shark, measuring 170 cm, was caught by a trawler in the Golfe de Gabès (Bradaï et al., 2002). On May 26, 1998, a 250 cm female shark was caught in a tuna trap off Sidi Daoud. The reported weight was about 225 kg and photographic documentation of the shark exists (Fergusson, 1998a). Fergusson (1996) reported this record based on a personal communication he received from Christian Capapé.

On April 23, 2001, a male shark was caught in a tuna trap 2.5–2.6 km off Sidi Daoud. On that afternoon, the wind was light and the sea was calm. The tuna trap was in water 35 m deep. This shark was reported measuring 490 cm in length and weighing 2200 kg. The stomach contained the vertebral column of a tuna. A tooth from this specimen has been preserved by Philippe Laulhe (Walidd Maamouri, pers. comm.; Philippe Laulhe, pers. comm.). Four days later, on April 27, another white shark was caught in the same tuna trap 2.5–2.6 km off Sidi Daoud. Conditions were similar to those four days earlier. The shark was reported measuring 520 cm in length and weighing 2250 kg. The original source reported the shark as a male, but the photo clearly shows it was a female. The stomach of the shark contained a large sea turtle. Like the

This 480 cm female white shark was caught in a tuna cage on October 17, 2009, off Mahdia, Tunisia (photograph by Mohamed Dahmouni).

previous shark, a tooth from the specimen was preserved by Philippe Laulhe. Photographic documentation of this specimen exists (De Maddalena and Reckel, 2003; Philippe Laulhe, pers. comm.; Walid Maamouri, pers. comm.).

On February 26, 2004, a large shark was caught by a tuna boat using a purse seine in the Golfe de Gabès. After the capture, which occurred in the afternoon, the shark was towed to Sfax harbor, 38 km from the capture location. This shark measured 587 cm and its weight was estimated at over 2000 kg. This large female shark was pregnant with four embryos, three females and one male (one male and one female in the right uterus and two females in the left uterus). The abdomen of the embryos was considerably distended by the yolk mass. The embryos were removed from the uteri and transferred to the Laboratoire d'ichtyologie de l'Institut des sciences et technologies de la mer de Sfax, where they were measured and weighed. They ranged in length from 132 cm to 135 cm and in weight from 27.65 kg to 31.50 kg. The stomach of the female shark was empty. Saidi et al. (2005) were unable to thoroughly examine the other organs because as soon the embryos were removed the mother was rapidly cut up and sold. One of the embryos was preserved at the Institut national des sciences et technologies de la mer, Centre de Sfax (cat. no. INSTM/LAM 01). Photographic documentation of this shark exists (Saidi et al., 2005). On April, 20, 2006, a young female was caught off Aras Dizra. It was captured in a net of the fishing vessel belonging to Mohamed Chelbaïa and weighed approximately 350 kg gutted. The shark was examined at the auction in Zarzis by Samuel P. Iglésias of the Muséum National d'Histoire Naturelle in Paris, who also collected a tissue sample for DNA analysis (tissu: BPS-0539) (Iglésias, 2001).

In 2008 a male white shark was caught 60 km off Chebba. Video documentation of this shark exists and author De Maddalena has estimated its total length at about 420 cm from the video. It was reported to weigh about 1500 kg. In October 2009, off Madia during the transfer of tuna from one tuna cage to another, an incident caused serious injuries to a number of tuna and the death of some of them. Three days later, the divers that checked the tuna cage saw a female great white shark swimming close to the tuna cage. As a results all the divers panicked and refused to dive outside the cage. Two days later, on October 17, 2009, the tuna cage staff discovered that the white shark had broken through the netting at the bottom and had entered the tuna cage. The shark started to attack the tuna, especially the weak individuals that were already injured. The divers attempted to release the shark, but one of them hit the animal near the eye with a knife, causing massive blood loss. The next day (October 18, 2009) the shark was found dead on the cage bottom. It measured 480 cm and weighed 1426 kg. The stomach contained remains of tuna and albatrosses. The shark showed three previous bites from other white sharks. Photographic and video documentation of this shark exists (Mohamed Dahmouni, pers. comm.).

Algeria

Two reports from Algeria occurred in the 1800s. The first was before 1865, when a 223 cm-long male shark was caught off Algeria. This shark is probably the same one that Guichenot donated to the Museum National d'Histoire Naturelle in Paris, France, which is mentioned by Dieuzeide and Novella (1953). Also, this is probably the white shark that Bernard Séret reported measuring 234 cm and is preserved as a taxidermied specimen (cat. no. MNHN a-9695) (De Maddalena, 2006b, 2007; Dieuzeide and Novella, 1953; Dumeril, 1865; Bernard Séret, pers. comm.; Marco Zuffa, pers. comm.). The second shark was caught in 1882 off Alger. Its jaws are preserved at the Zoologisk Museum in Copenhagen, Denmark (without cat. no.) (De Maddalena, 2006b, 2007; Jørgen Nielsen, pers. comm.). Before 1999, a small shark measuring 150 cm was caught in an unknown location off Algeria (Mohamed Hamdine, pers. comm.). Before 2006, another shark was caught in an unknown location off Algeria. Its jaws are preserved at the Museum National d'Histoire Naturelle in Paris, France (cat. no. MNHN a-9922) (De Maddalena, 2006b, 2007; Bernard Séret, pers. comm.; Marco Zuffa, pers. comm.). On an unknown date, perhaps in 2008, a large female white shark was caught by a trawler 200 m off Ghazaouet. Video documentation of this shark exists.

Morocco

Before 1926, a white shark was reported in the Golfe de Chafarinas, but it is unclear if it was captured (Fergusson, 1996). In March 1980, a 555 cm white shark was caught off Tetouan. Photographic documentation of this shark exists (Marco Zuffa, pers. comm.).

Unknown Locations

Adriatic Sea

The great white shark data bank has a few records from the Adriatic Sea without exact locations. A set of jaws from a shark caught on an unknown date from the Adriatic Sea is preserved at the Museo di Storia Naturale in Trieste, Italy (without cat. no.). This shark may have been caught between 1872 and 1905, when the Maritime government issued rewards for every great white shark captured (sharks were brought to the museum for identification) (De Maddalena, 2000c). From April 1872 to July 1882, 21 sharks caught in the northeastern Adriatic Sea ranged from 146 cm to 530 cm in length. They were presented to the Museo di Storia Naturale in Trieste, Italy, for species identification, which was required in order to obtain the monetary reward issued by the Imperial Maritime Austrian government for every great white shark captured. Of these sharks, seven measured over 400 cm. Surely some of these sharks, or possibly all, have already been accounted for among the Italian and Croatian records from the Adriatic Sea (Marchesetti, 1884). On June 17, 1879, two sharks were caught in the Adriatic Sea and presented to the Central Maritime government in order to obtain the reward money (Ninni, 1912; Perugia, 1881). Before September 1891, a huge white shark that measured 1005 cm and weighed 4000 kg was caught in the Adriatic Sea (Anonymous, 1891; Ellis and McCosker, 1991).

In 1901, a 500 cm shark was caught in the eastern Adriatic Sea (Barac, 1901; Soldo and Jardas, 2002), and before 1919, a 438 cm female shark was caught in the Adriatic Sea. The stomach contained the boots of a seaman from the Austro-Hungarian navy. This shark is preserved as a taxidermied skin-mount at the Naturhistorische Museum in Vienna, Austria (cat. no. NMW-95054). The time of capture was determined by the fact that the Austro-Hungarian navy was dissolved in 1918 (De Maddalena, 2006b, 2007; Helmut Wellendorf, pers. comm.). In May 1947, a white shark that measured 300 cm and weighed 300 kg was recorded in the eastern Adriatic Sea, but it is unclear if it was captured (Anonymous, 1947; Soldo and Jardas, 2002). In 1969, a white shark was recorded in the central part of the eastern Adriatic Sea, but it is unclear if it was captured (Morovic, 1969). Another shark was recorded before 1969 in the Adriatic Sea. The stomach contained a raincoat, three coats (or two, depending on the source) and a car license plate (Gianturco, 1978, Lineaweaver III and Backus, 1969).

Mediterranean Sea

There are a few records from the Mediterranean that do not specify the exact location of the encounter. Possibly in the 16th century, a shark estimated at over 500 cm in length was caught that had a seaman and two tunas in its stomach (Brunnich, 1768). In 1758, a sailor fell overboard from a frigate and was swallowed by a shark. The captain fired a gun at the shark, and the animal regurgitated the man, still alive. The shark was harpooned, dried, and presented to the sailor, who exhibited it throughout Europe. It was said to be 600 cm-long (GSAF). The species was not reported, but the only 6 m shark able to swallow a whole man in the Mediterranean is the great white shark. A basking shark *Cetorhinus maximus* has a mouth large enough to do the same, but such a mistake by this gentle filter-feeding giant has never been reported in the literature and it seems highly unlikely that a basking shark would swallow a man. A white shark was caught in 1984 on a longline at an unspecified location of the Mediterranean Sea (Rey et al., 1986). A 429 cm white shark was caught during the period from 1998 to 2001 on a longline at an unspecified location of the eastern Mediterranean Sea (Megalofonou et al., 2005).

A 420 cm shark caught before 1881 was preserved as a taxidermied skin-mount at the Museo di Storia Naturale in Genova, Italy (De Maddalena, 2000c; Moreau, 1881). It is possible that one of two sets of jaws preserved at the Istituto di Zoologia in Genova, which are labeled as from an unknown locality, may belong to this specimen. At the Koninklijk Belgisch Instituut voor Natuurwetenschappen in Brussels, Belgium, a chondrocranium, jaws and vertebrae from a shark caught before May 1, 1900, are preserved (cat. no. 1385 C) (De Maddalena, 2006b, 2007; Georges Lenglet, pers. comm.). A set of jaws from a large white shark is preserved at the Centre de Conservation et d'Etude des Collections, Musée des Confluences, in Lyon. The museum catalogue lacks any other data for this specimen (Didier Berthet, pers. comm.; De Maddalena and Zuffa, 2008).

Specimens Without a Capture Location

In addition to the cases mentioned above, there are also many cases that were recorded without any capture location. It is likely that some of these specimens were from the Mediterranean. Before 1881, a white shark that was reported measuring 548 cm and weighing 1814 kg was caught, possibly in the Mediterranean (Moreau, 1881). In the second half of May 2000, an estimated 500 cm shark was caught, most likely in the Mediterranean. The shark was seen at the fish market in Milan by Ruggero Ilgrande, who described the teeth as large and a few centimeters long. The shark was cut in pieces and sold for human consumption. However, there is some doubt over the exact identity of the species (De Maddalena and Piscitelli, 2001; Ruggero Ilgrande, pers. comm.).

In many of the natural history museums in Europe, there are many preserved white shark items that are without a capture location. Some of these specimens were likely captured in the Mediterranean. Not only is the capture location missing, but these specimens do not have a capture date as well. One of the preserved items is a head (cat. no. MR 1996–204), housed at the Muséum Requien in Avignon (Evelyne Crégut, pers. comm.; De Maddalena and Zuffa, 2008). A lower jaw (cat. no. OSTEO-130) and a set of jaws (numéro d'inventaire: OSTEO-220) are preserved at the Muséum d'Histoire Naturelle in Nice (De Maddalena and Zuffa, 2008; Olivier Gerriet, pers. comm.). A set of jaws (numéro d'inventaire: MHNGr.OS.47) is preserved at the Muséum d'Histoire Naturelle in Grenoble (De Maddalena, 2006) and another set of jaws (numéro d'inventaire: MHN LM 2005.7.1) is preserved at the Musée Vert–Musée d'Histoire Naturelle in Mans (De Maddalena and Zuffa, 2008; Nicolas Morel, pers. comm.). A set of jaws (numéro d'inventaire: 20) is preserved at the Muséum d'Histoire Naturelle de Nîmes (De Maddalena, 2006). A 500 cm taxidermied specimen (numéro d'inventaire: MNHNa-4669) and four sets of jaws (numéros d'inventaire: MNHN ab-0002, MNHN ab-0003, MNHN ab-0004, MNHN ab-0143) are preserved at the Muséum National d'Histoire Naturelle in Paris (De Maddalena, 2006). Another set of jaws (numéro d'inventaire: 2006.00.713) is preserved at the Muséum d'Histoire Naturelle in Valence (De Maddalena and Zuffa, 2008; Pascale Soleil, pers. comm.).

There are other specimens that have been added to this list of museum specimens without a capture location. A 383 cm specimen was preserved at the Muséum National d'Histoire Naturelle in Paris (Dumeril, 1865). A set of jaws was preserved at the Muséum Requien in Avignon, but seems now to be missing (Evelyne Crégut, pers. comm.; De Maddalena and Zuffa, 2008; Marco Zuffa, pers. comm.).

Before 1640–1660 a shark estimated at over 450 cm was caught, possibly in the Mediterranean. The set of jaws that was preserved belongs to the Biblioteca Ambrosiana in Milano (without cat. no.) but is presently at the Museo di Storia Naturale in Milano. The jaws were originally preserved in the ancient Settala Museum in Milano, Italy. This museum was owned by a nobleman named Manfredo Settala (1600–1680), who collected about 3000 items, including scientific instruments, minerals, animals, etnographical and archeological objects, coins, books and various oddities. Settala put his collection on exhibit in Milan, becoming one of the largest "Wunderkammer" of Europe (De Michele et al., 1983; Tavernari, 1976). Two ancient catalogues of the Settala Museum cite the large set of white shark jaws that are preserved at the Biblioteca Ambrosiana in Milan. The first citation appeared in an illustrated catalogue of the Settala Museum (Anonymous, 1640–1660), where the jaws were portrayed in an accurate watercolor figure and described as "Head, or mouth of one of the largest white sharks with three hundred teeth, that sucking cut a man in a thousand parts."

The second citation appeared in Terzago and Scarabelli (1677). Unfortunately, neither capture date nor location is reported in either catalogue. The collection was originally assembled by Manfredo Settala's father, Ludovico Settala (1552–1633), and later enriched by items collected during Manfredo's travels in Italy, Turkey, Cyprus and

This 438 cm female white shark, caught before 1919 in the Adriatic Sea, is preserved as a taxidermied skin-mount at the Naturhistorische Museum in Vienna, Austria (cat. no. NMW-95054) (photograph by Helmut Wellendorf, courtesy Naturhistorisches Museum Wien).

Africa, as well as by exchanges he made with other collectors. Thus it is impossible to determine the provenance of the jaws. Immediately after Manfredo Settala's death in 1680, the collection was partially divided, in spite of his express wishes that the entire collection be acquired by the Biblioteca Ambrosiana in Milan. It was not until the years 1751–1755 that what remained of the Settala collection was finally transported to the Biblioteca Ambrosiana (Navoni, 2000). In 1970, the natural history section of the Settala collection was loaned to the Natural History Museum of Milan, and a part, including the white shark jaws, was displayed in its first hall as part of a long-term exhibition of the Settala Museum. Recently, the exhibition was dismantled and the jaws were relocated in the Natural History Museum with the rest of the natural history portion of the Settala collection. Thanks to the Italian Ichthyological Society, the jaws were again exhibited to the public for a short period in October 2004 in the exhibition "*Alla ricerca del grande squalo bianco*" [In search of the great white shark], at the Sala Civica dei Disciplini, Castenedolo, Italy (De Maddalena, 2005). The jaws were then returned to the Natural History Museum of Milan, where they await return to the Biblioteca Ambrosiana for exhibition in the Settala Hall, together with the entire Settala collection. This is the oldest white shark item preserved in Europe (De Maddalena, 2006a).

The lower jaw of a juvenile specimen caught in the 18th century, belonging to the collection assembled by Italian naturalist Lazzaro Spallanzani, is preserved at the Musei

Civici in Reggio-Emilia, Italy (cat. no. 14-4527) (De Maddalena, 2000c). The upper jaw of a juvenile specimen caught in the 19th century is preserved at the Museo di Storia Naturale in Venezia, Italy (cat. no. 4860 Ist.V.S.L.A. 133) (De Maddalena, 2000c; Mizzan, 1994; Trois, 1900). The jaws of a specimen caught in 1868 or between 1872 and 1878 (cat. no. Z6431), the brain of a specimen caught in 1868 (cat. no. Z1113), an eye of a specimen caught in 1876 (cat. no. Z1194), and the heart of a specimen caught in 1876 or 1878 (cat. no. Z1244) are preserved at the Museo Zoologico in Napoli, Italy (Maio et al., 2005; De Maddalena, 2006b; De Maddalena, 2007; Nicola Maio, pers. comm.). A brain (cat. no. 111-208) and a set of jaws (cat. no. 111-167) from specimens caught before 1870 and a cranium from a specimen caught before 1707 (without cat. no.) are preserved at the Museo di Anatomia Comparata in Rome, Italy (the set of jaws with cat. no. 111-167 and the brain were originally preserved in the ancient Pontificium Romanum Archigymnasium in Rome, Italy, while the cranium without cat. no. was originally preserved in the ancient Museum Kircherianum) (Bonanni, 1707; Ernesto Capanna, pers. comm.; De Maddalena, 2000c, 2006b, 2007). Another set of jaws is preserved at the Museo di Storia Naturale in Genova, Italy (without cat. no.) (De Maddalena, 1997, 2000c). A cranium is preserved at the Museo di Anatomia Comparata in Torino, but it belongs to the Museo Regionale di Scienze Naturali in Torino, Italy (cat. no. 612) (De Maddalena, 2000c, 2002). Part of an upper jaw is preserved at the Museo di Storia Naturale e della Strumentazione Scientifica in Modena, Italy (cat. no. -049/91) (De Maddalena, 2000c). A cranium from a specimen caught probably before the 20th century is preserved at Museo di Storia Naturale e del Territorio in Calci, Italy (without cat. no.) (De Maddalena, 2000c).

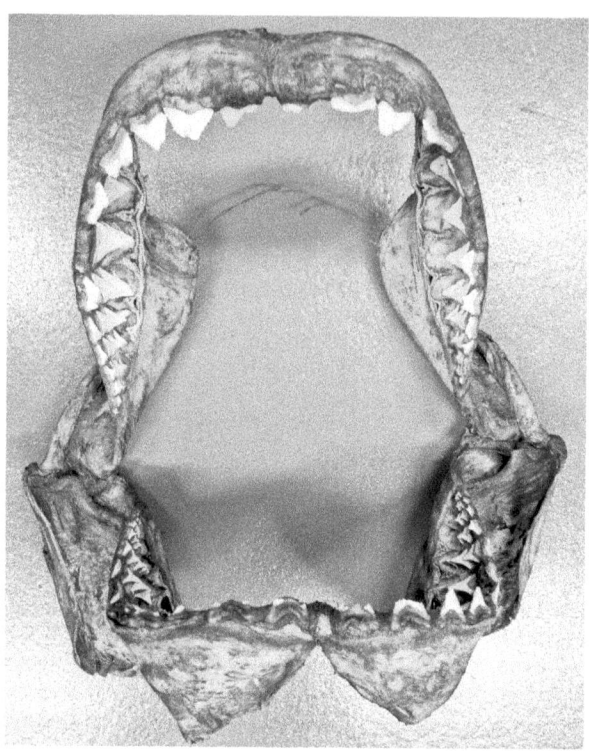

This set of jaws from a white shark caught in an unknown location of the Mediterranean Sea is preserved at the Centre de Conservation et d'Etude des Collections, Musée des Confluences de Lyon, France (courtesy Centre de Conservation et d'Etude des Collections, Musée des Confluences de Lyon).

A set of jaws is preserved at the Museo di Storia Naturale in Livorno (without cat. no.) (De Maddalena, 1997, 2000c). A set of jaws is preserved at the Museo di Storia Naturale — Sezione di Zoologia "La Specola" of the University of Florence, in Florence, Italy (cat. no. 6361) (De Maddalena, 2000c; Vanni, 1992). There are two sets of jaws preserved at the Istituto di Zoologia in Genova (without cat. no.). Those two sets may be from two of the three specimens caught in the following locations with the associated years: Isola Piana (summer 1879), Mediterranean (before 1881), Tyrrhenian (before 1881) (De Maddalena, 2000c). A set of jaws is preserved at the Museo di Storia Naturale in Ferrara, Italy (De Maddalena, 2000c). The Oceanographic Museum in Monaco has 25 preserved teeth from a white shark (Michèle Bruni, pers. comm.; De Maddalena, 2006b, 2007). A set of jaws from a specimen caught before 1863 is preserved at the

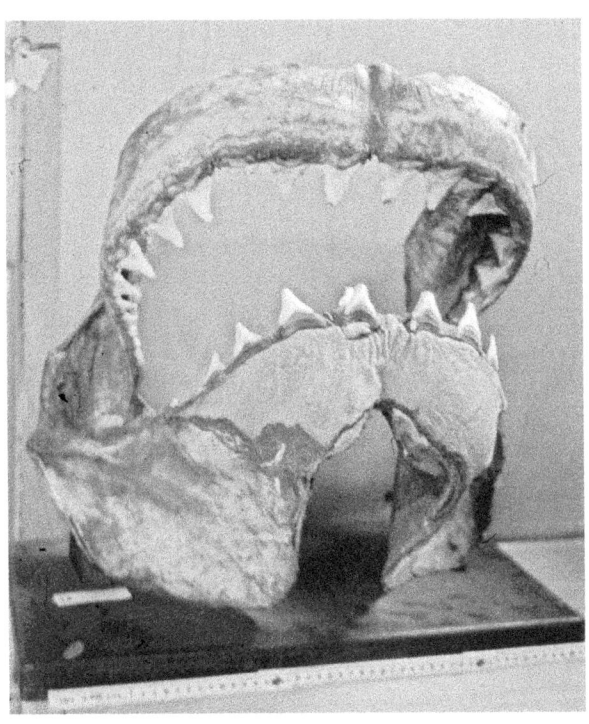

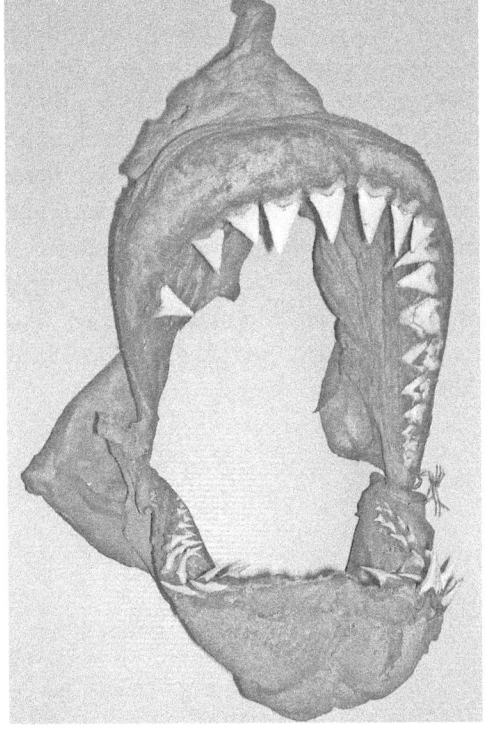

Top: This set of jaws from a white shark caught before 1640–1660, originally preserved in the ancient Settala Museum in Milan, is preserved at the Biblioteca Ambrosiana in Milan, Italy (without cat. no.) (photograph by Alessandro De Maddalena, courtesy Biblioteca Ambrosiana, Milano). *Right:* This cranium from a white shark caught before 1707, originally preserved in the ancient Museum Kircherianum in Roma, is preserved at the Museo di Anatomia Comparata in Roma, Italy (without cat. no.) (photograph by Ernesto Capanna, courtesy Museo di Anatomia Comparata di Roma, Italy).

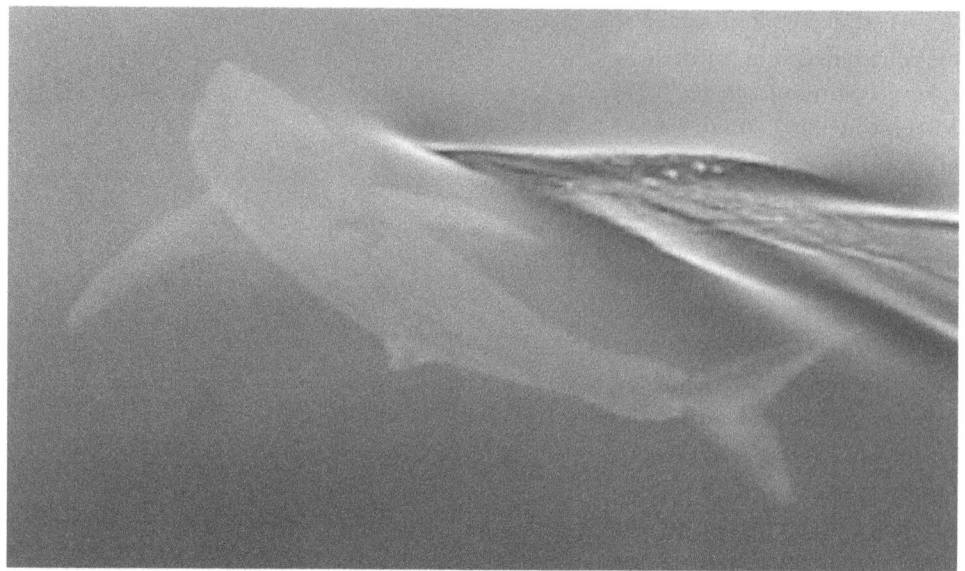

This estimated 400–430 cm female white shark was encountered on June 21, 2011, off Capraia, Italy (photographed by Marco Tarantino).

Musei Civici in Reggio-Emilia, Italy (no.cat. 11–4?4) (De Maddalena, 2000c; Moreau, 1881). A set of jaws from a shark caught before September 18, 1902, was bought at the fish market in Florence and is currently preserved at the British Museum of Natural History in London, United Kingdom (cat. no. BM(NH) 1905.12.2.2) (Oliver Crimmen, pers. comm.; De Maddalena, 2006b, 2007). A 223 cm-long white shark caught before 1909 was preserved as a taxidermied skin-mount at the Museo di Storia Naturale in Genova, Italy (Condorelli and Perrando, 1909; De Maddalena, 2000c) but this item is currently missing. Also from a shark caught before 1909 is a set of jaws preserved at the Museo di Storia Naturale in Palermo, Italy. Condorelli and Perrando (1909) wrote that this jaw set was old. According to Mojetta et al. (1997), the shark was caught off La Spezia in 1880. These jaws are likely one of the five jaw sets preserved at the museum with cat. nos. An-108, An-115, An-128, An-145, and An-80 (De Maddalena, 2006b, 2007).

3

Summary Tables

Date	Location	Length (cm)	Weight (kg)	Sex	Remarks	Sources
SPAIN — Continental Spain						
January 1878	Between Islas Columbretes and Castellón SPAIN	Ca. 500	2300		Captured in a net.	Barrull and Mate (2001), Lozano-Rey (1928), Perez-Arcas (1878)
Ca. 1920	Melilla SPAIN	150			Captured	Fergusson (1996)
Ca. 1927	Premiá de Mar SPAIN				Captured	F.X. Viñals Moncusi (pers. comm.)
August 10, 1946	Tabarca SPAIN	Ca. 550 (est.)	1790		Captured in a tuna trap. Stomach contained a tuna	http://lafogueradetabarca.blogspot.com
1962	Castellón	500	1500		Capture. Stomach contained a tuna and a dolphin.	Asensi (1977), Barrull (1993–1994) F.X. Viñals Moncusi (pers. comm.)
March (or Jun) 18, 1983 or 1986	Tarifa SPAIN	Ca. 350 (est.)			Nonfatal attack on windsurfer L. Pérez-Diaz Doubtful case.	J. Barrull (pers. comm.), GSAF, F.X. Viñals Moncusi (pers. comm.)
November	Tossa de Mar SPAIN	475	1000	M	Stranded. Jaws preserved at the Centro de Recuperación de Animales Marinos Fundación CRAM in Premiá de Mar, Spain.	Barrull (1993, 1993–1994), Barrull et al. (1999) Barrull and Mate (2001), F.X. Viñals Moncusi (pers. comm.)
Before 1878	Puerto de Mazarrón SPAIN				Captured? Doubtful case.	Perez-Areas (1878) F.X. Viñals Moncusi (pers. comm.)
Before 1878	Valencia SPAIN	500			Captured.	Perez-Areas (1878)
Before December 16, 1912	Vilassar de Mar SPAIN	471	1000		Captured in a tuna trap. Two teeth preserved at the Museu de Zoologia in Barcelona, Spain (cat. no. MZB-82-5316 and MZB-82-5317)	Barrull (1993–1994) Barrull et al. (1999) Barrull and Mate (2000)

Date	Location	Length (cm)	Weight (kg)	Sex	Remarks	Sources
Before 1926	Tarragona SPAIN				Captured. Doubtful case.	Barrull et al. (1999) De Buen (1926), Fergusson (1996), F.X. Viñals Moncusi (pers. comm.)
Before 1928	Vinaroz SPAIN				Captured. Jaws presumably preserved at the Museo Nacional de Ciencias Naturales in Madrid, Spain.	Lozano-Rey (1928)
Before 1999	Alboran SPAIN				Captured on a longline.	F.X. Viñals Moncusi (pers. comm.)
SPAIN — Islas Baleares						
Before 1868	Menorca, Islas Baleares SPAIN				Captured?	Barcelo and Combis (1868), Fergusson (1996)
1920s	Majorca, Islas Baleares SPAIN	Ca. 390 (est.)			Captured in a tuna trap.	Morey et al. (2003)
1920s	Majorca, Islas Baleares SPAIN		2000		Captured in a tuna trap.	Morey et al. (2003)
Winter 1920s	Majorca, Islas Baleares SPAIN	700			Captured in a tuna trap.	Morey et al. (2003)
September 3, 1927	Majorca, Islas Baleares SPAIN	700			Captured in a tuna trap.	Morey et al. (2003)
Winter 1935	Majorca, Islas Baleares SPAIN	700			Captured in a tuna trap.	Morey et al. (2003)
Winter 1940s	Majorca, Islas Baleares SPAIN	>400	800		Captured in a tuna trap.	Morey et al. (2003)
Winter 1940s	Majorca, Islas Baleares SPAIN	>400	1000		Captured in a tuna trap.	Morey et al. (2003)
February 1, 1942	Majorca, Islas Baleares SPAIN	400	800		Captured in a tuna trap. Stomach contained a common thresher shark.	Morey et al. (2003), Ramis (1988)
February 12, 1944	Majorca, Islas Baleares SPAIN	535	1350		Captured in a tuna trap.	Morey et al. (2003)

3 — Summary Tables

Date	Location	Length (cm)	Weight (kg)	Sex	Remarks	Sources
March 1962	Majorca, Islas Baleares SPAIN	350	500		Captured in a tuna trap. Photographic documentation exists.	Morey et al. (2003)
March 1962	Majorca, Islas Baleares SPAIN	300	300		Captured in a tuna trap. Photographic documentation exists.	Morey et al. (2003)
Winter 1963	Majorca, Islas Baleares SPAIN	500	500		Captured in a tuna trap.	Morey et al. (2003)
December 26, 1963	Majorca, Islas Baleares SPAIN	615	2200	F	Captured in a tuna trap. Photographic documentation exists.	Morey et al. (2003)
Winter 1965	Majorca, Islas Baleares SPAIN	700	1000		Captured in a tuna trap.	Morey et al. (2003)
January 1964	Majorca, Islas Baleares SPAIN	535 (510 TL — est.)	1400	F	Captured in a tuna trap. Photographic documentation exists.	Morey et al. (2003)
1966	Majorca, Islas Baleares SPAIN	1100		F	Captured in a tuna trap.	Morey et al. (2003)
1966	Majorca, Islas Baleares SPAIN	1200		F	Captured in a tuna trap.	Morey et al. (2003)
January 1967	Majorca, Islas Baleares SPAIN	550 (est.)	1700	F	Captured in a tuna trap. Photographic documentation exists.	Morey et al. (2003)
October 1 1967	Majorca, Islas Baleares SPAIN	515 (450 TL — est.)	1700	M	Captured in a tuna trap. Photographic documentation exists.	Morey et al. (2003)
1967	Majorca, Islas Baleares SPAIN	535	1350	F	Captured in a tuna trap by Francisco Pérez. Stomach contained a dolphin and four large tunas. Photographic documentation exists.	Morey et al. (2003), Ramis (1988)

Date	Location	Length (cm)	Weight (kg)	Sex	Remarks	Sources
February 1969	Majorca, Islas Baleares SPAIN	550 (535 TL — est.)	2000	F	Captured in a tuna trap. Photographic documentation exists.	Morey et al. (2003)
February 1969	Majorca, Islas Baleares SPAIN	Ca. 550 (est.)	1250		Captured in a tuna trap. Photographic documentation exists.	Morey et al. (2003)
February 1969	Majorca, Islas Baleares SPAIN		1000		Captured in a tuna trap.	Morey et al. (2003)
March 1969	Majorca, Islas Baleares SPAIN	800 (620 TL — est.)	2500	F	Captured in a tuna trap. Photographic documentation exists.	Morey et al. (2003)
January 1970	Majorca, Islas Baleares SPAIN		1350		Captured in a tuna trap.	Morey et al. (2003)
1972	Majorca, Islas Baleares SPAIN		2500		Captured in a tuna trap.	Morey et al. (2003)
February 5, 1976	Majorca, Islas Baleares SPAIN	615 (610 TL — est.)	2500	F	Captured in a tuna trap. Photographic documentation exists. Stomach contained a large manta, or ray.	Morey et al. (2003), Ramis (1988), M. Zuffa (pers. comm.)
July 1992	Andraitx, Majorca, Islas Baleares SPAIN	500			Two specimens captured. Doubtful case.	Barrull and Mate (2001), Fergusson (1996)
July 5, 1997	Islas Baleares SPAIN				Carcass of an 8 m sperm whale stranded bearing white shark bites.	Morey et al. (2003)
March 21, 1998	Islas Baleares SPAIN				Carcass of a 2.2 m striped dolphin stranded bearing white shark bites in the urogenital region.	Morey et al. (2003)

3 — Summary Tables

Date	Location	Length (cm)	Weight (kg)	Sex	Remarks	Sources
June 13, 1998	Islas Baleares SPAIN				Carcass of a loggerhead sea turtle stranded bearing white shark bites on the pelvic flippers, carapace and plastron.	Morey et al. (2003)
September 8, 1998	Islas Baleares SPAIN				Carcass of a loggerhead sea turtle stranded bearing white shark bites on the plastron.	Morey et al. (2003)
July 23, 2000	Islas Baleares SPAIN				Carcass of a 3.3 m bottlenose dolphin stranded bearing white shark bites in the urogenital region.	Morey et al. (2003)
August 9, 2000	Islas Baleares SPAIN				Carcass of a 3.5 m bottlenose dolphin stranded bearing white shark bites in the urogenital region.	Morey et al. (2003)
August 2000	Cabrera, Islas Baleares SPAIN	Ca. 400 (est.)			Encountered. Doubtful case.	Barrull and Mate (2001)
March 3, 2001	Islas Baleares SPAIN				Carcass of a Risso's dolphin stranded bearing white shark bites.	Morey et al. (2003)
April 19, 2001	Islas Baleares SPAIN				Carcass of a 1.9 m striped dolphin stranded bearing white shark bites in the urogenital region.	Morey et al. (2003)
			FRANCE — Continental France			
Middle Ages	Aix-en-Provence? FRANCE	670			Captured. Stomach contained a man wearing armor. Doubtful case.	Gianturco (1978), Smith (1833)
Before 1554	Nice FRANCE				Captured. Stomach contained a man wearing armor.	Rondelet (1554)

Date	Location	Length (cm)	Weight (kg)	Sex	Remarks	Sources
Before 1554	Marseille FRANCE				Captured. Stomach contained a man wearing armor.	Gessner (1560), Rondelet (1554)
18th Century	Between Cassis and La Ciotat FRANCE				Captured. Stomach contained two tuna and an entire dressed man.	Cazeils (1998)
18th Century	Antibes FRANCE				Attack on a bathing seaman.	Cazeils (1998)
18th Century	Nice FRANCE				Fatal attack on a child.	Cazeils (1998)
18th Century	Île Sainte-Marguerite FRANCE				Captured. Stomach contained an entire horse.	Cazeils (1998)
19th Century	FRANCE?			F	Captured. Doubtful case.	Lacépède (1839)
May 1861	Sète FRANCE				Captured. Row of teeth preserved at the Muséum National d'Histoire Naturelle in Paris, France (cat. no. MNHN ab-0195).	De Maddalena (2006b, 2007), B. Séret (pers. comm.)
August 1875	Sète FRANCE	Ca. 400	600 600		Captured.	Moreau (1881)
1876	Sète FRANCE	242			Captured.	Moreau (1881)
1888	Le Grau-du-Roi FRANCE	250		M	Captured. Preserved as a taxidermied skin-mount at the Muséum d'Histoire Naturelle de Nîmes, France (cat. no. BAC 132).	De Maddalena (2006b, 2007), G. Gory (pers. comm.)
1889	La Seyne-sur-Mer FRANCE				Captured. Four teeth preserved at the Muséum National d'Histoire Naturelle in Paris, France (cat. no. MNHN ab-0185).	De Maddalena (2006b, 2007), B. Séret (pers. comm.)

3 — Summary Tables

Date	Location	Length (cm)	Weight (kg)	Sex	Remarks	Sources
October 1889	Le Brusc FRANCE	400			Captured in a net. Stomach contained a porpoise and the legs and pelvis of a man.	Cazeils (1998), Dujardin (1890), Moreau (1892)
Before 1893	Toulon FRANCE				Captured?	Carus (1893)
Before 1893	Nice FRANCE				Some sharks captured.	Carus (1893), Moreau (1881)
Ca. 1898	Var FRANCE	550	Ca. 2000		Captured in a net. Stomach contained a porpoise, three tuna and other fishes.	Bonomi (1898)
Before 1909	Camargue FRANCE				Captured. Jaws preserved at the Musée Océanographique de Monaco, Monaco (cat. no. MOM-P0I-4254).	Bruni and Würtz (2002), De Maddalena (2006b, 2007), M. Bruni (pers. comm.)
Before 1909	Martigues FRANCE				Captured. Jaws preserved at the Musée Océanographique de Monaco, Monaco (cat. no. MOM-P0I-4253, old cat. no. 911).	M. Bruni (pers. comm.), Bruni and Würtz (2002), De Maddalena (2006b, 2007), Roule (1912)
October 11, 1910	Le Grau-du-Roi FRANCE				Captured. Three teeth preserved at the Muséum d'Histoire Naturelle de Nîmes, France (cat. no. 6).	De Maddalena (2006b, 2007), G. Gory (pers. comm.)
1912	Six-Four-les Plages FRANCE				Captured.	G. Altman (pers. comm.)
October 15, 1925	Marseille FRANCE	600	1500	F	Captured by A. Rouard. Photographic documentation exists.	De Maddalena and Révelart (2008), De Maddalena and Zuffa (2008), P. Summonti (pers. comm.).
Summer 1935	Pointe de l'Espiguette FRANCE	500	1600 (gutted)		Captured. Stomach contained three whole tunas.	Perrier (1938)

Date	Location	Length (cm)	Weight (kg)	Sex	Remarks	Sources
Summer 1935	Languedoc FRANCE				Captured.	Perrier (1938)
Summer 1935	Languedoc FRANCE				Several sightings.	Perrier (1938)
Ca. 1940	Le Grau-du-Roi FRANCE	800			Captured. Stuffed and exhibited throughout the Provence region.	Granier (1964)
1943	Between Palavas and Grau-du-Roi, FRANCE				Captured.	Granier (1964)
1946	Tra Palavas and Grau-du-Roi FRANCE				Captured.	Granier (1964)
Ca. 1950	Toulon FRANCE				Attack on two fishermen.	Touret (1992)
1956	Between the calanques of Niolon and Figuerolles FRANCE		1800		Captured in a drift net ("thonaille") by Mr. Scotto and other fishermen. Stomach contained the remains of several dolphins, large tunas and a monk seal pup. Photographic documentation exists.	Damonte (1993), Perosino (1963)
October 13, 1956	Maguelone FRANCE	589 TOT	2000	F	Captured in a trawl by M. Antoine Ferrignon and other fishermen. Stomach contained the remains of two dolphins measuring about 180 cm. Preserved as a cast and the original fins and teeth in the Musée cantonal de Zoologie in Lausanne, Switzerland (without cat. no.).	de Beaumont (1957), De Maddalena et al. (2003), Quignard et al. (1962)
1956	Between Palavas and Grau-du-Roi FRANCE				Captured.	Granier (1964)

Date	Location	Length (cm)	Weight (kg)	Sex	Remarks	Sources
Before 1965	Marseille FRANCE				Attack on a worker that fell into the waters of the harbor.	Hemingway and Devlin (1965)
December 4, 1986	Îlot Tiboulen, Îles du Friol FRANCE	Ca. 500–600 (est.)			Encountered by two divers, Marc Ferrand and Alain Alziari.	A. Alziari (pers. comm.), Anonymous (1986), De Maddalena and Révelart (2008), De Maddalena and Zuffa (2008)
1989	Marseille FRANCE	>400 est.			Sighted by fishermen following a school of tuna. Photographic documentation exists.	Touret (1992)
Ca. 1990	Valras FRANCE				Encountered by M. Arthur while he was fishing.	G. Oliver (pers. comm.)
January 9, 1991	Sète FRANCE	600 (591 TOT — est)	Ca. 2000	F	Captured in a trawl by Jean Licciardi and other fishermen. Stomach contained the remains of four dolphins measuring from 80 to 100 cm, and two swordfish. Photographic documentation exists.	Anonymous (1991b), De Maddalena et al. (2001), Quignard and Raibaut (1993)
Ca. 1993	Between Nice and the Corsica FRANCE				Encountered by some sportfishermen.	G. Altman (pers. comm.)
1998	Cap d'Antibes FRANCE				Nonfatal attack on a diver.	P Brocchi (pers. comm.)
October 6, 2001	Between Cap Croisette and Cap Caveaux FRANCE	Ca. 630–680 (est.)			Encountered by Claude Wagner and nine divers.	J. Attard (pers. comm.), De Maddalena and Herber (2002), C. Wagner (pers. comm.)
June 20, 2002	Cap Ferrat FRANCE	Ca. 400 ca. (est.)			Encountered by diver Christian Gelpi and two other divers.	C. Gelpi (pers. comm.)

Date	Location	Length (cm)	Weight (kg)	Sex	Remarks	Sources
FRANCE — Corsica						
September 14, 1984	Îlot des Moines, Roccapina, Corsica FRANCE	527	Ca. 1500		Captured in a trammel net by Antoine-Jean Gianetti, Jean-Baptiste Gianetti, Thomas Duval and Antoine Duval. Stomach contained a 60 kg dolphin. Photographic documentation exists.	Anonymous (1984), R. Miniconi (pers. comm.), Miniconi (1987, 1994)
October 7, 1989	Mouth of Solenzara River, Corsica FRANCE				Encountered by Pietro Gambini. Some doubt exists over the exact identity of the species.	F. Serena (pers. comm.)
February 1995	Capo di Feno, Corsica FRANCE	Ca. 600 (est.)			Encountered by fisherman Antoine Deriu and another witness.	P.H. Weber (pers. comm.)
August 2008	Golfe d'Ajaccio, Corsica FRANCE	Ca. 400–500 (est.)			Encountered by cetacean observer Pierre–Henri Weber.	P.H. Weber (pers. comm.)
August 2009	Golfe de Valinco, Corsica FRANCE	Ca. 400–500 (est.)			Encountered by seaman Christophe Recco and three other witnesses. Film documentation exists.	P.H. Weber (pers. comm.)
December 2009	Between Sagone and Toulon, Corsica FRANCE	>600 (est.)			Encountered by boat captain Baptiste Bacchiolelli and another witness. Film documentation exists.	P.H. Weber (pers. comm.)
April 2010	Golfe d'Ajaccio, Corsica FRANCE	Ca. 350–450 (est.)			Encountered by fisherman Antoine Deriu and 10 other witnesses.	P.H. Weber (pers. comm.)
ITALY — Ligurian Coast						
Before 1846	Santa Margherita Ligure ITALY	640			Captured.	Sassi (1846), Fergusson (1996)

3 — *Summary Tables*

Date	Location	Length (cm)	Weight (kg)	Sex	Remarks	Sources
1876	Portofino ITALY	446		M	Captured. It was preserved as a taxidermied skin-mount at the Museo di Storia Naturale in Milan, Italy (cat. no. 2142).	Tortonese (1938)
1879	Viareggio ITALY	308 TLn		F	Captured. Preserved as a taxidermied skin-mount at the Museo di Storia Naturale — Sezione di Zoologia "La Specola" of the University of Florence, in Florence, Italy (cat. no. 5983). Photographic documentation exists.	De Maddalena (1997, 2000c), Giglioli (1880), Vanni (1992)
1880	La Spezia ITALY				Captured. Jaws preserved at the Museo di Storia Naturale in Palermo, Italy (likely one of the five jaw sets with cat. nos. An-108, An-115, An-128, An-145, and An-80).	De Maddalena (2006b, 2007), Mojetta et al. (1997)
January 1883	Portofino ITALY	430 TLn		F	Captured. Preserved as a taxidermied skin-mount at the Museo di Storia Naturale e della Strumentazione Scientifica in Modena, Italy (cat. no. -045/91). Stomach contained dogs, cats, molluscs, a pair of pants, a pair of boots and pieces of canvas. Photographic documentation exists.	Condorelli and Perrando (1909), De Maddalena (1999, 2000c)

Date	Location	Length (cm)	Weight (kg)	Sex	Remarks	Sources
December 10, 1891	Monterosso ITALY	600	600 (gutted)	F	Captured. Jaws preserved at the Museo di Storia Naturale — Sezione di Zoologia "La Specola" of the University of Florence, in Florence, Italy (cat. no. 6032). Photographic documentation exists.	De Maddalena (1997, 1998, 2000c), Tortonese (1956), Vanni (1992)
September 14, 1896	Capo Vado ITALY				Some sharks observed scavenging on the carcass of an 18 m whale, likely a fin whale.	Parona (1896)
October 19, 1896	Genova ITALY				Some sharks observed scavenging on the carcass of a 21 m whale, likely a fin whale. Some doubt exists over the exact identity of the species.	Parona (1896)
May-June 1909	Celle Ligure ITALY	350	350		Captured.	A. Mojetta (pers. comm.)
September 1909	Sestri Ponente ITALY				Captured.	A. Mojetta (pers. comm.)
1914	Portofino ITALY	Ca. 450 (475 TOT — est.)			Captured. Photographic documentation exists.	Mojetta et al. (1997), M. Zuffa (pers. comm.)
July 23, 1926	Varazze ITALY	Ca. 600–700 (est.)			Unprovoked fatal attack on bather Augusto Cesellato, witnessed by Luigi Baldi.	Anonymous (1926a, 1926b)
End of July 1928	Viareggio ITALY				Provoked nonfatal attack on the boat of Mr. Bagolini. Some doubt exists over the exact identity of the species.	Gianturco (1978)

3 — Summary Tables

Date	Location	Length (cm)	Weight (kg)	Sex	Remarks	Sources
1929	Portofino ITALY	414		M	Captured. Preserved as a taxidermied skin-mount at the Museo di Storia Naturale in Genova, Italy (cat. no. C.E. 27517).	De Maddalena (1999, 2000c), Mojetta et al. (1997), Tortonese (1956)
July 3, 1935	Riva Trigoso ITALY				Captured. Jaws preserved at the Museo di Storia Naturale in Genova, Italy (cat. no. C.E.32695).	De Maddalena (1999, 2000c)
June 1953–1955	Santa Margherita Ligure ITALY	80			Captured by hook and line. Some doubt exists over the exact identity of the species.	A. Mojetta (pers. comm.), Mojetta et al. (1997)
May 16, 1954	Camogli ITALY	700 or 520–550 (est.)?	1400 or 2000	F	Captured in a tuna trap. Accompanied by a larger specimen. Photographic documentation exists.	Cappelletti (1989b), Fergusson (1996), Melegari (1973), Tortonese (1965)
March 25, 1956	Genova ITALY	Ca. 700 (est.)?			Attack on a boat.	Mojetta et al. (1997)
Summer 1961	Camogli ITALY	Ca. 500 (est.)			Encountered by spearfishing diver F.K.	Roghi (1962)
Ca. 1970	Punta Chiappa ITALY				Captured in a tuna trap.	Albertarelli (1990)
June 15, 1983	Riomaggiore ITALY	Ca. 300 (est.)			Encountered by scuba diver Roberto Piaviali.	GSAF
July 30, 1991	Portofino ITALY	Ca. 400 (est.)			Unprovoked non-fatal attack on the kayak of Ivana Iacaccia.	Fergusson (1996), Graffione (1991), A.L. Recchi (pers. comm.)
October 24, 1999	Punta Chiappa ITALY	Ca. 600–700 (est.)			Sighted by Salvatore Cilona from a coast guard helicopter searching for the wreckage of a small plane that crashed into the sea.	Danker (2001), De Maddalena (2000a), Preve (1999)

Date	Location	Length (cm)	Weight (kg)	Sex	Remarks	Sources
		ITALY — Sardinia				
1777	Sardinia ITALY				Captured.	A. Mojetta (pers. comm.)
Summer 1879	Isola Piana, Sardinia ITALY	405			Captured in a tuna trap. It was preserved as a taxidermied skin-mount at the Museo di Zoologia in Genova, Italy.	De Maddalena (2000c), Parona (1919), M. Zuffa (pers. comm.)
May 22, 1882	Isola Piana, Sardinia ITALY	>300			Captured in a tuna trap.	Parona (1919)
Ca. 1970	Mal di Ventre Sardinia ITALY				Unprovoked non-fatal attack on the boat of a sportfisherman.	M. Cottiglia (pers. comm.)
April 1971	Capo Testa, Sardinia ITALY	>400		M	Captured in a tuna trap. Jaws preserved and displayed at the restaurant Le Bocche di Bonifacio in Capo Testa, Sardinia, Italy.	Fergusson (1996), M. Zuffa (pers. comm.)
Late spring or early summer 1975	Secca della Frasca, Capo della Frasea, Sardinia ITALY				Captured in a gill net.	M. Cottiglia (pers. comm.)
August 14, 1976	Santa Caterina di Pittinurri, Sardinia ITALY	470 or 400 or ca. 500 (est.)	Ca. 1100	M	Captured in a tuna trap. Stomach contained the remains of sea turtles and some common dentex. Photographic documentation exists.	Fergusson (1996, 1998c), Mojetta et al. (1997)
1976	Santa Teresa di Gallura, Sardinia ITALY				Captured. Photographic documentation exists.	Anonymous (1976)

3 — Summary Tables

Date	Location	Length (cm)	Weight (kg)	Sex	Remarks	Sources
June 1977	Capo Testa, Sardinia ITALY	580		M	Captured in a tuna trap by Salvatore Murru. Jaws preserved and displayed at the restaurant Le Bocche di Bonifacio in Capo Testa, Sardinia, Italy. Photographic documentation exists.	Mojetta et al. (1997), M. Zuffa (pers. comm.)
April 1978	Santa Teresa di Gallura, Capo Testa, Sardinia ITALY	Ca. 200			Captured in a tuna trap. Photographic documentation exists.	Malatesta (1987)
August 24, 1991	Santa Teresa di Gallura, Sardinia ITALY				Sighted.	A. Mojetta (pers. comm.)
June 1997	Golfo del Leone, Sardinia ITALY				Encountered by the Italian navy. Domenico Marcianò witnessed the shark swallow 2–3 seagulls and a 50×50 cm box containing garbage.	D. Marcianò (pers. comm.)
September 1999	Alghero, Sardinia ITALY	Ca. 400 (est.)			Encountered by surfer Tore Manca.	De Maddalena (2000a), T. Manca (pers. comm.)
August 29, 2001	Torre delle Stelle, Sardinia ITALY	Ca. 400 (est.)			Encountered by sportfisherman Mauro Cottiglia.	M. Cottiglia (pers. comm.)
May 13, 2007	Carloforte, Sardinia ITALY	Ca. 300 (est.)			Entered a tuna trap but succeeded in freeing itself and fled the area.	Froldi (2007)
2009	Alghero, Sardinia ITALY				Several sightings by fishermen.	M. Colombo (pers. comm.).
September 1, 2009, and August 29, 2010	Alghero, Sardinia ITALY				Four white shark tooth fragments found by snorkelers on the seafloor.	M. Colombo (pers. comm.).

Date	Location	Length (cm)	Weight (kg)	Sex	Remarks	Sources
		ITALY — Tyrrhenian Coast				
October 1666	Livorno ITALY				Captured. Head acquired by the Medici in Florence, who gave it to Niels Stensen for study.	Stenone (1667)
June 6, 1721	Napoli ITALY	527	1425	F	Fatal attack on a bather collecting seafood. Captured by placing a looped rope around its head. Stomach contained many fish and human remains, including a half cranium with hair, part of a vertebral column, and ribs.	Ricciardi (1721)
1822	Napoli ITALY	150		F	Captured. Preserved as a taxidermied skin-mount at the Senckenberg Forschungsinstitut und Naturmuseum in Frankfurt am Main, Germany (without cat. no.).	De Maddalena (2006b, 2007) F. Krupp (pers. comm.), Müller and Henle (1838)
December 1876	Canale di Piombino ITALY	1400			Captured.	Lawley (1881)
May 7, 1883	Scilla ITALY	1336			Captured with a harpoon. Stomach contained many human bones (including a femur or a tibia and a cranium with hair) and human clothes, including a pair of boots, three shoes, three socks, a wool bodice, and a pair of pants.	Anonymous (1833)

3 — Summary Tables

Date	Location	Length (cm)	Weight (kg)	Sex	Remarks	Sources
1886	Piombino ITALY	800–1000			Captured in a tuna trap. Stomach contained an entire human corpse sewn into a sailcloth with cast-iron weights, clearly a seaman buried at sea. Three teeth preserved.	V. Biagi (pers. comm.), Biagi (1989, 1995), Mojetta et al. (1997)
June 5, 1898	Portoferraio, Isola d'Elba ITALY				Captured.	Brian (1906)
November 23, 1910	Marciana Marina, Isola d'Elba ITALY				Observed feeding on a dying whale, likely a fin whale.	Damiani (1911), Mojetta et al. (1997)
1924	Between Scilla and Bagnara ITALY	1200			Captured. Some doubt exists over the exact identity of the species.	Gianturco (1978)
Mid June 1924	Procida ITALY	Ca. 600 (507 TOT — est.)	Ca. 1200		Captured in a tuna trap. Photographic documentation exists.	Anonymous (1924), De Maddalena et al. (2001)
1930	Pizzo Calabro ITALY	1400			Captured in a tuna trap.	Comune di Pizzo Calabro (1991), Mojetta et al. (1997)
1933	Tirreno ITALY				Two sharks captured. Jaws were said to be preserved at the Museo di Storia Naturale in Genova, Italy.	De Maddalena (2000c), Mojetta et al. (1997)
1933–1934	Port'Ercole ITALY	800			Captured by hook and line.	Storai et al. (2000)
August 12, 1938	Enfola, Isola d'Elba ITALY	Ca. 600 (597–613 TOT — est.)	1800	F?	Captured in a tuna trap by the Ridi brothers. Stomach contained two dolphins. Photographic documentation exists.	Anonymous (1938), De Maddalena (1999), De Maddalena et al. (2001)

Date	Location	Length (cm)	Weight (kg)	Sex	Remarks	Sources
1940	Isola d'Elba ITALY				Captured in a tuna trap. Doubtful case.	A. Mojetta (pers. comm.)
1949–1950	San Vincenzo ITALY				Nonfatal attack on a boat.	V. Biagi (pers. comm.)
1952–1953	Enfola, Isola d'Elba ITALY	Ca. 400–500 (est.)			Sighted.	Storai et al. (2000)
May 1, 1950	Sprizze, Isola d'Elba ITALY				Nonfatal attack on the boat of fisherman Remo Adriani. Some doubt exists over the exact identity of the species.	F. Serena (pers. comm.)
September 1956	Secca del Faro, Circeo ITALY	420	>600		Nonfatal attack on scuba diver Goffredo Lombardo. Captured by hook and line and killed with rifle shots.	Lombardo (1960)
Ca. 1960	Isola d'Elba ITALY				Fatal attack on three people (a mother, father and their six-year-old son). Doubtful case.	Hemingway and Devlin (1965)
Early summer 1960	Secca del Quadro, Circeo ITALY	Ca. 700–800 (est.)			Encountered by spearfishing diver Maurizio Sarra.	Carletti (1973), Giudici and Fino (1989)
August 18, 1960	Rio Marina, Isola d'Elba ITALY>	Ca. 600	1000	M	Captured in a trammel net operated by Gennaro Chiocca, Alfonso Chiocca, Giacomo Chiocca and Giuseppe Pinotti. Stomach contained two 15 kg dolphins. Photographic documentation exists.	G. Chiocca (pers. comm.), Chiocca (1990), De Maddalena (1999)
1961	Circeo ITALY				Sighted.	Carletti (1973)
Summer 1962	Circeo ITALY				Sighted.	Carletti (1973)

3 — Summary Tables

Date	Location	Length (cm)	Weight (kg)	Sex	Remarks	Sources
End of August 1962	Circeo ITALY				Encountered by spearfishing divers Carlo and Raimondo Bucher.	Carletti (1973), Giudici and Fino (1989)
End of August 1962	Circeo ITALY				Unprovoked non-fatal attack on diver Dante Matacchione.	Carletti (1973), Giudici and Fino (1989)
September 2, 1962	Secca del Quadro, Circeo ITALY				Unprovoked fatal attack on spearfishing diver Maurizio Sarra.	Carletti (1973), Gianturco (1978), Gilioli (1989), Giudici and Fino (1989), Marini (1989)
July 20, 1964	Canale di Montecristo	>900	1900		Observed scavenging on the carcass of a fin whale. Captured by fisherman Sirio Scotto, who put a loop around its head. Stomach contained a seagull, a small dolphin in two parts, a black wig, a broomstick and pieces of the whale. Photographic documentation exists.	Rosselli (...)
November 18–23, 1964	Secca del Quadro, Circeo ITALY	320 & Ca. 650 (est.)	380	M &F	Male and female encountered by spearfishing divers Gino Felicioni and Cesare Polidori on November 18, 1964. Probably one of the same sharks encountered on November 19, 1964. Female harpooned but succeeded in freeing itself and fled the area; male captured by hook and line, harpooned and killed with rifle shots. on November 22 and 23, 1964. Photographic documentation exists.	Maltini (1965)

Date	Location	Length (cm)	Weight (kg)	Sex	Remarks	Sources
July 22, 1967	Canale dell'Isola del Giglio ITALY	>900			Captured in a net. Stomach contained dolphins and large pelagic fish.	Anonymous (1967), Storai et al. (2000)
March 1968	Civitavecchia ITALY				Sighted. Some doubt exists over the exact identity of the species.	Pirino and Usai (1991)
May 5, 1968	Marciana Marina, Isola d'Elba ITALY				Captured in a net.	F. Serena (pers. comm.)
Before 1971	Secca di Capo d'Omo, Argentario ITALY				Captured in a trammel net. Stomach contained two dolphins, a sea turtle and several kg of fish.	Betti (1971)
September 5, 1973	Secche di Vada ITALY				Encountered by diver Dino Podestà. Some doubt exists over the exact identity of the species.	F. Serena (pers. comm.)
September 15, 1975	Secche di Vada ITALY				Encountered by diver Dino Podestà.	F. Serena (pers. comm.)
May 21, 1976	Civitavecchia ITALY				Sighted.	A. Mojetta (pers. comm.)
October 26, 1977	Secche della Meloria ITALY	700 (est.)			Sighted?	Anonymous (1979), Mojetta et al. (1997)
1978	Near Rende ITALY	Ca. 500	800		Captured in a drift net. Photographic documentation exists.	De Maddalena (1997), M. Zuffa (pers. comm.)
September 16, 1978	Capo d'Anzio ITALY	Ca. 500 (est.)		F	Unprovoked non-fatal attack on diver Fabrizio Marini.	Marini (1989)
1982	Biodola, Isola d'Elba ITALY				Attack on the boat of fisherman Giuseppe Campodonico. Doubtful case.	Perfetti (1989), M. Zuffa (pers. comm.)

3 — Summary Tables

Date	Location	Length (cm)	Weight (kg)	Sex	Remarks	Sources
1982	Lacona, Isola d'Elba ITALY	Ca. 600 (est.)			Attack on the boat of Giovanni Vuoso. Doubtful case.	Perfetti (1989), M. Zuffa (pers. comm.)
July 1983 or 1984	Baratti ITALY	350 (est.) & Ca. 300 (est.)		F&F	Two sharks captured on a longline, witnessed by Gianfranco Della Rovere.	G. Della Rovere (pers. comm.)
December 3, 1984	Marciana Marina, Isola d'Elba ITALY				Attack on a boat.	F. Serena (pers. comm.)
1987	Marciana Marina, Isola d'Elba ITALY				Attack on the boat of Aniello Mattera and Giorgio Allori. Doubtful case.	Perfetti (1989), M. Zuffa (pers. comm.)
June 1987	Scilla ITALY				Scavenging on a captured swordfish.	Giudici and Fino (1989)
August 15, 1988	Procchio, Isola d'Elba ITALY				Encountered by the commander of a research ship. Some doubt exists over the exact identity of the species.	F. Serena (pers. comm.)
August 17, 1988	Maratea ITALY	Ca. 500 (est.)			Sighted.	A. Mojetta (pers. comm.), Perfetti (1989)
January 13, 1989	Golfo di Baratti ITALY				Observed preying. Some doubt exists over the exact identity of the species.	F. Serena pers. comm.)
February 2, 1989	Scoglio Stella, Golfo di Baratti ITALY	>600 (est.)			Unprovoked fatal attack on spearfishing diver Luciano Costanzo, witnessed by Gianluca Costanzo and Paolo Bader.	Albertarelli (1990), Bertuccelli (1989), Biagi (1989), Cappelletti (1989a), Giudici and Fino (1989), F. Serena (pers. comm.)
April 10, 1989	Canale di Piombino ITALY				Sighted.	Capuozzo (1989), A. Mojetta (pers. comm.)

Date	Location	Length (cm)	Weight (kg)	Sex	Remarks	Sources
May 18, 1989	Canale di Piombino ITALY				Sighted.	Anonymous (1989b), Mojetta et al. (1997)
June 6, 1989	Marinella di Sarzana ITALY	Ca. 300 (est.)			Unprovoked nonfatal attack on windsurfer Ezio Bocedi.	E. Bocedi (pers. comm.), GSAF
September 9, 1989	Bocca d'Arno ITALY	Ca. 400–500 (est.)			Encountered by sportfishermen Maurizio Bini, Tommaso Salone and Paolo Sbrana. Photographic documentation exists.	Conti (1990)
1990	Livorno ITALY				Sighted?	A. Mojetta (pers. comm.)
1990	Baratti ITALY				Sighted. Some doubt exists over the exact identity of the species.	Cappelletti (1990)
May 25, 1990	Favazzina ITALY	>1000 (est.)			Encountered by fishermen Agatino Billé and Giacomo Lisciotto.	Aricò (1990), A. Celona (pers. comm.)
May 15, 1990	Marciana Marina, Isola d'Elba ITALY				Sighted. Some doubt exists over the exact identity of the species.	F. Serena (pers. comm.)
Before 1991	Livorno ITALY	>400 (est.)			Observed pursuing a school of Atlantic bonito.	Pirino and Usai (1991)
September 8, 1998	Capo Vaticano ITALY	Ca. 600 (est.)			Scavenging on a captured swordfish.	Anonymous (1998), G. Spagnolo (pers. comm.)
December 27, 1998	Scoglio Stella, Golfo di Baratti ITALY	Ca. 700 (est.)			Encountered by sportfishermen Roberto Cheli and Luca Antonio and the coast guard. Photographic documentation exists.	De Sabata et al. (1999), Gasperetti (1998)
Before 2000	Tyrrhenian Sea ITALY				Captured. Jaws preserved at the Museo di Storia Naturale in Genova, Italy (cat. no. C.E. 31916).	De Maddalena (2000c, 2002)

3 — Summary Tables

Date	Location	Length (cm)	Weight (kg)	Sex	Remarks	Sources
July 22, 2000	Isola del Giglio ITALY	Ca. 200 (est.) & Ca. 200 (est.)	Ca. 120 (est.) & Ca. 120 (est.)		Two sharks encountered by sportfishermen Alessandro Vitturini, Vito Cofrancesco, Giancarlo Graziani, Mario Marini and Salvatore Meli.	V. Cofrancesco (pers. comm.), Porqueddu (2000)
February 11, 2001	Riva di Traiano ITALY	Ca. 650–700 (est.)			Encountered by sportfishermen Vito Cofrancesco, Alessandro Vitturini, Paolo Santi, Giancarlo Graziani and Luigi Signorelli. Observed scavenging on dead sardines, scorpionfish and squid. Photographic documentation exists.	V. Cofrancesco (pers. comm.), Fanelli (2001), A. Vitturini (pers. comm.)
October 7, 2002	Secche di Vada ITALY	Ca. 400–500 (est.)			Encountered by scuba divers Paolo Virnicchi, Paolo Spinelli, Franco Salomone and Alessandro Barlettani.	P. Virnicchi (pers. comm.), P. Spinelli (pers. comm.), A. Barlettani (pers. comm.)
June 10, 2006	Isola Gorgona ITALY	Ca. 500–700 (est.)			Encountered by sportfisherman Vincenzo Giudice.	V. Giudice (pers. comm.)
June 21, 2011	Isola Capraia ITALY	Ca. 400–430 (est.)		F	Encountered by researchers. Filmed documentation of this specimen exists.	C. Volpi (pers. comm.)
Before 1996	Livorno ITALY				Captured?	Fergusson (1996)
ITALY — Sicily						
Before 1864	Golfo di Catania, Sicily ITALY	Ca. 300			Captured.	Gemmellaro (1864)
Ca. 1880	Catania, Sicily ITALY	300			Captured.	Doderlein (1881)
Ca. 1881	Stretto di Messina, Sicily ITALY				Nonfatal attack on a boat of fishermen.	Doderlein (1881), Mojetta et al. (1997)

Date	Location	Length (cm)	Weight (kg)	Sex	Remarks	Sources
January 1889	Sferracavallo, Capo Gallo, Sicily ITALY				Captured. Jaws preserved at the Museo di Storia Naturale in Palermo, Italy (likely one of the five jaw sets with cat. nos. An-108, An-115, An-128, An-145, and An-80).	De Maddalena (2006b, 2007), Riggio (1894)
December 1892	Capo Gallo, Sicily ITALY	>400	1500	M	Captured. Jaws preserved at the Museo di Storia Naturale in Palermo, Italy (likely one of the five jaw sets with cat. nos. An-108, An-115, An-128, An-145, and An-80).	De Maddalena (2006b, 2007), Riggio (1894)
August 1893	Pace, Sicily ITALY	460 FOR	Ca. 1125		Captured by a fisherman with a harpoon.	Facciolà (1894)
Before 1894	Golfo di Catania, Sicily ITALY				Captured.	Riggio (1894)
1894	Golfo di Catania, Sicily ITALY				Captured?	A. Mojetta (pers. comm.)
March 24, 1894	Isola delle Femine, Sicily ITALY	Ca. 200			Captured. Jaws preserved at the Museo di Storia Naturale in Palermo, Italy (likely one of the five jaw sets with cat. nos. An-108, An-115, An-128, An-145, and An-80).	De Maddalena (2006b, 2007), Riggio (1894)

3 — Summary Tables

Date	Location	Length (cm)	Weight (kg)	Sex	Remarks	Sources
Between the end of 1908 and early 1909	Messina, Sicily ITALY				Stranded, but still alive. Stomach contained the leg of a woman, cleanly cut. The victim undoubtedly was drowned by the tsunami that occurred after an earthquake that took place on December 20, 1908.	Berdar and Riccobono (1986), Munthe (1928)
Before 1909	Golfo di Catania, Sicily ITALY	304			Captured.	Condorelli and Perrando (1909)
January 26, 1909	Augusta, Capo S.Croce, Sicily ITALY	450	800		Observed scavenging on the carcass of a dolphin. Captured. Stomach contained remains of at least three humans: a man, a woman and a six-year-old child. The victims undoubtedly were drowned by the tsunami that occurred after an earthquake that took place on December 20, 1908.	Condorelli and Perrando (1909)
1913	Torre Faro, Messina, Sicily ITALY				Captured. Stomach contained an estimated 300 kg tuna that was swallowed in two parts. Jaws preserved.	Celona (2002a)
August 27, 1934	Catania, Sicily ITALY	Ca. 400 (est.)			Unprovoked nonfatal attack on a boat full of fishermen.	Giudici and Fino (1989)
May 23, 1935	Stretto di Messina, Sicily ITALY	>500		M	Captured. Stomach contained two tuna that were 35–40 kg each.	Scordia (1935)

Date	Location	Length (cm)	Weight (kg)	Sex	Remarks	Sources
August 23, 1937	Marzamemi, Sicily ITALY	530	Ca. 1200	M	Captured with a harpoon. There were no contents in the stomach. Photographic documentation exists.	Celona (2002b)
1947	Scopello, Sicily ITALY	Ca. 900–1000 (est.)	3590		Captured in a tuna trap. Tooth preserved by fisherman Toti Lo Iacono. Photographic documentation exists.	G. Spagnolo (pers. comm.), M. Omodei (pers. comm.), M. Zuffa (pers. comm.)
May 1953	Favignana, Isole Egadi ITALY				Two specimens entered a tuna trap but succeeded in freeing themselves and fled the area.	G. Guarrasi (pers. comm.)
May 29, 1953	Favignana, Isole Egadi ITALY	550	>1700	F	Captured in a tuna trap, succeeded in breaking the nets and freeing itself, but was killed by the military with a machine gun. Stomach contained many large fish, including a 15 kg tuna, and garbage, including a 5 kg can.	Carletti (1973), G. Cataldo (pers. comm.), De Maddalena (2002), Giudici and Fino (1989), G. Guarrasi (pers. comm.), Jannuzzi (1962), S. Spataro (pers. comm.)
1960	Capo Passero, Sicily ITALY	>600			Captured in a tuna trap. Photographic documentation exists.	Carletti (1973)
Ca. between 1960 and 1990	Favignana or Isola la Formica, Isole Egadi ITALY				Entered a tuna trap but succeeded in freeing itself and fled the area.	N. Mineo (pers. comm.)
Ca. between 1960 and 1990	Favignana or Isola la Formica, Isole Egadi ITALY				Entered a tuna trap but succeeded in freeing itself and fled the area.	N. Mineo (pers. comm.)

3 — Summary Tables

Date	Location	Length (cm)	Weight (kg)	Sex	Remarks	Sources
Ca. between 1960 and 1990	Favignana or Isola la Formica, Isole Egadi ITALY				Entered a tuna trap but succeeded in freeing itself and fled the area.	N. Mineo (pers. comm.)
June 19, 1961	Ganzirri, Sicily ITALY	Ca. 640 (est.) (666 TOT — est.)	Ca. 1500 (est.)	F	Captured with a harpoon by Domenico Sorrenti. Stomach contained a large dolphin that was swallowed in two parts. Photographic documentation exists.	Celona (2002a), De Maddalena et al. (2001)
1964	Isola la Formica, Isole Egadi ITALY	Ca. 600			Captured in a tuna trap.	N. Mineo (pers. comm.)
January 1965	Marzamemi, Sicily ITALY	Ca. 500	>800		Captured, entangled in a cable that was used to anchor pots used to fish caramote prawns. Stomach contained bones and hairs, probably from a wolf or dog. Photographic documentation exists.	Celona (2002b), Marino (1965)
March 9, 1965	Ganzirri, Sicily ITALY	620? (560 TOT — est.)			Observed chasing a school of mullet. Captured with a harpoon by Donato Nicola. Stomach contained parts of loggerhead sea turtles, remains of bony fish and other unidentified remains. Jaws preserved in the Istituto di Idrobiologia of the University of Messina in Ganzirri, Italy. Photographic documentation exists. Mojetta et al.	Berdar and Riccobono (1986), Celona et al. (2001), Giudici and Fino (1989), Mojetta (1997)

Mediterranean Great White Sharks

Date	Location	Length (cm)	Weight (kg)	Sex	Remarks	Sources
Ca. March 9, 1965 (?)	Ganzirri, Sicily ITALY	>700?	Ca. 3000		Captured by Antonino Arena.	Berdar and Riccobono (1986), Giudici and Fino (1989), Mojetta et al. (1997)
June 1967	Stretto di Messina, Sicily ITALY	Ca. 400–500 (est.)			Provoked non-fatal attack on a boat full of fishermen.	Giudici and Fino (1989)
Ca. 1970	Siracusa, Sicily ITALY	400			Captured?	Touret (1992)
Ca. 1970	Capo Rasocolmo, Sicily ITALY				Encountered by two fishermen.	A. Celona (pers. comm.)
Before 1973	Lampedusa, Banco del Mezzogiorno, Isole Pelagie ITALY				Captured in a net.	Carletti (1973)
Before 1973	Lampedusa, Isole Pelagie ITALY				Captured.	Carletti (1973)
May 1974	Isola la Formica, Isole Egadi ITALY	620–640 (594 TOT — est.)	2400–2600 or 1500	F	Captured in a tuna trap. Stomach contained an entire goat, plastic bottles, plastic bags and possibly an entire dolphin weighing 80 kg. Photographic documentation exists.	Anonymous (1974), G. Cataldo (pers. comm.), De Maddalena (2002), De Maddalena et al. (2001), Fergusson (1996), G. Guarrasi (pers. comm.), N. Mineo (pers. comm.)
August 1977	Mazara del Vallo, Sicily ITALY	135		M	Captured in a net.	Fergusson (1996)
1978	Favignana, Isole Egadi ITALY		Ca. 1600		Captured in a tuna trap.	G. Cataldo (pers. comm.), De Maddalena (2002), S. Spataro (pers. comm.)
August 1978	Mazara del Vallo, Sicily ITALY	220			Captured in a net.	Fergusson (1996)
June 1979	Lampione, Isole Pelagie ITALY	Ca. 200			Captured. Photographic documentation exists.	Merlo (1979)

3 — Summary Tables

Date	Location	Length (cm)	Weight (kg)	Sex	Remarks	Sources
April 24, 1980	Favignana, Isole Egadi ITALY	580 & 540	Ca. 1600	F & M	Two sharks captured in a tuna trap. Female landed on April 24, 1980, and male landed on April 25, 1980. Female stomach contained a partial skeleton of a swordfish and an estimated 20 kg dolphin in two parts. Photographic documentation exists.	Bruno (1980), G. Cataldo (pers. comm.) De Maddalena (2002), Fergusson (1996), N. Mineo (pers. comm.)
May 1980–1982	Bonagìa, Sicily ITALY	Ca. 600	1030		Captured in a tuna trap.	De Maddalena (2002), N. Ravazza (pers. comm.), M. Solina (pers. comm.)
Before 1981	Sciacca, Sicily ITALY			F	Pregnant female carrying six 40 cm-long embryos captured. Some doubt exists over the exact identity of the species.	Di Milia (1981)
August 1981	Pantelleria ITALY	202		F	Captured in a net. Photographic documentation exists.	Fergusson (1996)
May 1982	Scopello, Sicily ITALY	Ca. 600–700	1300		Captured in a tuna trap. Stomach contained horns of a goat, a plastic container, a boat chain, an English Navy boot, and half of an estimated 1.5 m long striped dolphin.	M. Omodei (pers. comm.)
August 10, 1983	Mazara del Vallo, Sicily ITALY	155		M	Captured in a net.	Fergusson (1996)
August 11, 1983	Mazara del Vallo, Sicily ITALY	142		F	Captured in a net.	Fergusson (1996)
Around 1984	Scopello, Sicily ITALY	Ca. 500 (est.)			Captured in a tuna trap.	G. Spagnolo (pers. comm.).

Date	Location	Length (cm)	Weight (kg)	Sex	Remarks	Sources
June 25, 1985	Mazara del Vallo, Sicily ITALY	220			Captured in a net.	Fergusson (1996)
September 1985	Mazara del Vallo, Sicily ITALY	Ca. 138–158 (est.)			Captured. Jaws preserved.	Vacchi and Serena (1997)
Mid July (?) 1985–1986	Punta Secca, Sicily ITALY	Ca. 305–366 (est.)			Unprovoked non-fatal attack on snorkeler Neil Montoya.	De Maddalena (2002), N. Montoya (pers. comm.)
March 1986	Mazara del Vallo, Sicily ITALY	200	170		Captured in a net.	Travaglini (1986)
March 1986	Pantelleria ITALY	Ca. 500			Captured.	Mojetta et al. (1997)
May 8, 1987	Favignana, Isole Egadi ITALY	535	2000 or 2200–2500	F & M	Two sharks, a male and a female, caught in a tuna trap. The male succeeded in breaking the net and fled, the female did not escape and was caught. Stomach contained a 150–200 kg dolphin that was already dead before it was eaten, since it had a 3 m long rope around its tail. Fins preserved at the Camperia in Favignana, Italy. Photographic documentation exists.	G. Cataldo (pers. comm.), Curzi (1987), De Maddalena (2002), Fergusson (1996), Giudici and Fino (1989), Soletti (1987), S. Spataro (pers. comm.)
July 31, 1987	Ganzirri, Sicily ITALY				Carcass of a 250 cm bottlenose dolphin stranded bearing white shark bites on its lower surface closer to the tail. Some doubt exists over the exact identity of the species.	Centro Studi Cetacei (1988)
1989	Isole Egadi ITALY				Captured.	Fergusson (1996)

3 — Summary Tables

Date	Location	Length (cm)	Weight (kg)	Sex	Remarks	Sources
June 23, 1989	Ganzirri, Sicily ITALY	>500 (est.)			Encountered by Antonio Celona from a boat fishing for swordfish.	A. Celona (pers. comm.)
August 13, 1989	Isola Galera, Isole Egadi ITALY	Ca. 300 (est.)			Three sharks encountered by diver G. Bertin.	Fergusson (1996), Mojetta et al. (1997)
Ca. 1990	Palermo, Sicily ITALY	Ca. 400			Captured by Giuseppe Storniolo. Stomach contained the vertebral column of a swordfish, a dolphin that was eaten in 3–4 parts, an albacore eaten in two parts and some feathers.	A. Celona (pers. comm.)
August 25, 1990	Capo Granitola, Sicily ITALY	Ca. 520		M	Captured in a net. Stomach contained garbage, including metallic cans. Photographic documentation exists.	De Maddalena (2002), Fergusson (1996), Mojetta et al. (1997), M. Zuffa (pers. comm.)
June 1991	Mazara del Vallo, Sicily ITALY	400			Captured.	Fergusson (1996)
July 15–20, 1991	Banco di Pantelleria ITALY	490–500 (est.)		F	Encountered by spearfishing diver Riccardo Andreoli. Photographic documentation exists.	R. Andreoli (pers. comm.), Fergusson (1996)
1992–1993	Lampedusa, Banco di Mezzogiorno, Isole Pelagie ITALY				A few sharks encountered by some fishermen.	Fergusson (1996)
May 1993	Favignana, Isole Egadi ITALY				Scavenging on a 180–200 kg tuna.	Mojetta et al. (1997)
May 1993	Favignana, Isole Egadi ITALY	140			Captured in a tuna trap. Doubtful case.	Mojetta et al. (1997), S. Spataro (pers. comm.)
Summer 1993	Portopalo di Capo Passero, Sicily ITALY				Captured in a net.	Mojetta et al. (1997)

Date	Location	Length (cm)	Weight (kg)	Sex	Remarks	Sources
November 1993	Mazara del Vallo, Sicily ITALY	135			Captured. Jaws preserved.	Fergusson (1996)
June 1995–1996	Strombolicchio, Isole Eolie ITALY	Ca. 500 (est.)			Encountered by diver Angelo Pistorio and other divers. Photographic documentation exists but it seems to be lost.	De Maddalena (2002), A. Pistorio (pers. comm.)
1997	Canale di Sicilia, Sicily ITALY	Ca. 200			Captured in a net.	Storai et al. (2000)
Summer 1997	Between Marinella Selinunte and Capo Granitola, Sicily ITALY	Ca. 300 (est.)			Several sightings by fishermen.	Capitaneria di Porto di Mazara del Vallo
August 14, 1998	Lampedusa, Isole Pelagie ITALY	150			Captured by Nicola Costa. Photographic documentation exists.	De Sabata et al. (1999)
October 19, 1999	Banco "Chitarro," Marsala, Sicily ITALY	Ca. 400–500 (est.)			Encountered by spearfishing diver Giovanni Bosco.	G. Bosco (pers. comm.)
November 2, 1999	Augusta, Sicily ITALY	Ca. 500 (est.)			Encountered by spearfishing diver Giovanni Zito.	Nania (1999)
May 14, 2001	Ganzirri, Sicily ITALY	Ca. 100			Sighted by fisherman Agatino Billé.	A. Celona (pers. comm.)
June 29, 2002	Ganzirri, Sicily ITALY	Ca. 400 (est.)			Sighted by fishermen.	A. Celona (pers. comm.)
May 5, 2006	Lampedusa, Isole Pelagie ITALY	>400 (est.)			A 350 cm bottle nose dolphin showing two fresh white shark bites on the dorsal region.	Celona et al. (2006)
July 12, 2006	Lampedusa, Isole Pelagie ITALY	Ca. 150 (est.)			Encountered by Gabriella La Manna and other biologists from the Ricerche Delfini CTS of Lampedusa. Photographic documentation exists.	Ufficio Stampa CTS

3 — Summary Tables

Date	Location	Length (cm)	Weight (kg)	Sex	Remarks	Sources
September 22, 2007	Banco di Pantelleria ITALY	Ca. 500 (est.)			Encountered by spearfishing skin-diver Alberto Zaccagni.	A. Zaccagni (pers. comm.)
ITALY — Ionian Coast						
September 18, 1979	Gallipoli ITALY	620	2700 or Ca. 1700	M	Captured in a trammel net by Pompeo Alessandrelli. Stomach contained a pair of shoes, a washing machine drum, a doll, a ham bone, one or more 1 kg meat cans still sealed. Photographic documentation exists.	R. Basso (pers. comm.), Piccinno and Piccinno (1979)
May 4, 1988	Bova Marina ITALY				Carcass of a 310 cm bottlenose dolphin stranded bearing a white shark bite in the urogenital region.	Centro Studi Cetacei (1989), M. Zuffa (pers. comm.)
Mid June 1989	Capo Spartivento ITALY	550			Captured in a driftnet. Stomach contained two 120 cm dolphins.	Centro Studi Cetacei (1991)
Ca. 2000	Punta Ristola ITALY				White shark tooth found by skin diver on the seafloor.	L. Di Giosa (pers. comm.)
Unknown date	Marina di Salve ITALY	Ca. 450		M	Captured. Photographic documentation exists.	
ITALY — Adriatic Coast						
September 16, 1823	Adriatic ITALY	490 TLn		F	Captured. Preserved as a skin-mount in the Museo Zoologico in Padova, Italy (cat. no. P25E).	Canestrini (1874), Condorelli and Perrando (1909), De Maddalena (2000c, 2002, 2006b, 2007), P. Nicolosi (pers. comm.)
1823	Adriatic? ITALY				Captured. Jaws preserved in the Museo di Anatomia Comparata in Bologna, Italy (cat. no. Alessandrini 811).	De Maddalena (2006b, 2007), D. Minelli (pers. comm.), M. Zuffa (pers. comm.)

Date	Location	Length (cm)	Weight (kg)	Sex	Remarks	Sources
Spring 1827	Adriatic ITALY				Captured. Jaws preserved at the Museo di Anatomia Comparata in Bologna, Italy (cat. no. Alessandrini 1216).	De Maddalena (2000b, 2000c), D. Minelli (pers. comm.), Mojetta et al. (1997), M. Zuffa (pers. comm.)
Early February 1839	Civitanova ITALY	Ca. 600 (602 TL — est.)	1814	F	Captured or stranded. Part of the skeleton, including the chondrocranium, the jaws and part of the vertebral column are preserved at the Museo di Anatomia Comparata of Rome, Italy (cat. no. 111-95).	Bonaparte (1839), Metaxà (1839), Vinciguerra (1890), Condorelli and Perrando (1909), De Maddalena (1997, 1998, 2000c)
September 1, 1868	Trieste ITALY				Fatal attack.	Radovanovi (1965), Soldo and Jardas (2002)
April 19, 1872	Grado ITALY	300			Captured.	Brusina (1888)
1873	Trieste ITALY	460		M	Captured.	Doderlein (1881), Graeffe (1886)
May 12, 1877	Adriatic ITALY				Captured.	Ninni (1912), Perugia (1881)
May 12, 1877	Adriatic ITALY				Captured.	Ninni (1912), Perugia (1881)
November 5, 1879	Grado ITALY	250			Captured.	Ninni (1912), Perugia (1881)
1880	Golfo di Trieste ITALY	460			Captured. It was preserved as a taxidermied skin-mount at the Museo di Storia Naturale in Venezia, Italy.	Mojetta et al. (1997), Ninni (1912)
Before 1881	Golfo di Venezia ITALY	490			Captured.	Carus (1893), Doderlein (1881)
September 14, 1885	Santa Croce di Trieste ITALY	400			Captured.	Brusina (1888)

3 — Summary Tables

Date	Location	Length (cm)	Weight (kg)	Sex	Remarks	Sources
1902	Trieste ITALY	375 TLn		M	Captured. Preserved as a taxidermied skin-mount at the Museo di Storia Naturale in Venezia, Italy (cat. no. 2039).	De Maddalena (2000b, 2000c), Mizzan (1994)
June 1908	Golfo di Trieste ITALY	1400			Captured by fisherman Stelio Candela.	Arrassich (1994)
Ca. 1945	Pescara ITALY	Ca. 600			Captured by fisherman Vittorio Pomante.	Cugini and De Maddalena (2003)
July 7, 1961	Riccione ITALY	Ca. 450 (est.)			Unprovoked nonfatal attack on spearfishing diver Manfred Gregor, witnessed by Heinz Schmid and Herbert Russauer.	Baldridge (1973), GSAF
June 7, 1978	Golfo di Venezia ITALY	Ca. 500 (est.)			Sighted by researchers Luigi Alberotanza and Luigi Cavaleri from the research platform Acqua alta. Regurgitated the remains of a bottlenose dolphin. Photographic documentation exists.	L. Alberotanza (pers. comm.), Albertarelli (1990), Beltrame (1983), L. Cavaleri (pers. comm.), De Maddalena (1999, 2000b), Gilioli (1989)
June 1978	Caorle ITALY				Sighted.	L. Alberotanza (pers. comm.), De Maddalena (2000b)

Date	Location	Length (cm)	Weight (kg)	Sex	Remarks	Sources
End of September 1986	Adriatic ITALY	Ca. 600 (est.)			Several sightings. Nonfatal attack on a fishing boat. Snatched a whole crate of pilchards from the hand of a fisherman. Sighted on September 20 by the captain of a hydrofoil along the Rimini–Yugoslavia route. Sighted on September 23 off Rimini near the oil-platform Antonella. Sighted off Pesaro, near the oil-platform Basil. Sighted by sportfishermen Gabriele Bartoletti and Stefano Dragoni.	Anonymous (1986), De Maddalena (2000b), Gilioli (1989), Giudici and Fino (1989), Marini (1989), Martelli (1989)
August– September 1987	Pesaro ITALY	>600 (est.)			Sighted.	Cardellini (1987), Mojetta et al. (1997)
May 1988	Numana ITALY	Ca. 450 (est.)			Encountered by sportfisherman Fausto Fioretti. Photographic documentation exists.	M. Marconi (pers. comm.), De Maddalena (1997, 1999, 2000b)
September 9, 1988	Porto Barricata ITALY	>550 (est.)			Encountered by sportfishermen. Photographic documentation exists.	Mojetta et al. (1997), M. Zuffa (pers. comm.)
September 1989	Pesaro ITALY	>500 (est.)			Sighted. Photographic documentation exists.	Fergusson (1996)
End of August 1991	Between Pesaro and Cattolica ITALY	>600 (est.)			Encountered by fisherman Armando Marzi and other fishermen.	Barone (1999)
December 17, 1991	Ancona ITALY	210	180		Captured in a net.	Fergusson (1996)

3 — Summary Tables

Date	Location	Length (cm)	Weight (kg)	Sex	Remarks	Sources
Before 1992	Northern Adriatic ITALY				Two sharks captured in nets.	Anonymous (1992)
Mid March 1992	Termoli ITALY	230	Ca. 200	F	Four or five sharks captured.	Anonymous (1992), Fergusson (1996)
1993	Isole Tremiti ITALY	Ca. 450 (est.)			Encountered by fisherman Nino. Some doubt exists over the exact identity of the species.	G. Cugini (pers. comm.)
September 15, 1996	Brindisi ITALY	400–500 (est.)			Encountered by spearfishing diver Leonardo Leone De Castris. Photographic documentation exists.	De Maddalena and Herber (2002), L. Leone De Castris (pers. comm.)
August 27, 1998	Senigallia ITALY	Ca. 500–600 (est.)	Ca. 1200		Encountered by sportfisherman Stefano Catalani. Scavenging on a captured common thresher shark. Film documentation exists.	Imarisio (1998), Montefiori (1998)
September 26, 1999	Giulianova ITALY	Ca. 600 (est.)			Encountered by fisherman Elvio Mazzagufo. Scavenging on a captured tuna. Photographic documentation exists.	De Maddalena (2000b), Fiaccarini (1999a, 1999b), Graziosi (1999)
September – early October 1999	Giulianova ITALY	Ca. 700 (est.)			Encountered by sportfisherman Antonio Bombini. Film documentation exists.	Barone (1999)
April 29, 2000	Piave Vecchia ITALY	Ca. 500 (est.)			Encountered by sportfishermen.	L. Mizzan (pers. comm.)
April 13, 2001	Torre Testa Rossa ITALY	Ca. 500–600 (est.)			Encountered by spearfishing divers Teddy Sciurti and Paolo D'Ambrosio.	De Maddalena and Herber (2002), T. Sciurti (pers. comm.)

Date	Location	Length (cm)	Weight (kg)	Sex	Remarks	Sources
July 30, 2001	Falconara ITALY	Ca. 550–600 (est.)			Encountered by fishermen Tersilio and Giacomo Longhi. Photographic documentation exists.	G. Longhi (pers. comm.)
September 9, 2002	Porto San Giorgio ITALY	Ca. 700–800 (est.)			Encountered by sportfishermen Glauco Micheli and Roberto Boracci. Photographic documentation exists.	G. Micheli (pers. comm.)
Before 2005	Foce del Po ITALY				Sighted. Photographic documentation exists.	M. Zuffa (pers. comm.)
ITALY — Unknown location						
1881	ITALY				Captured. Jaws preserved at the Museo di Storia Naturale e della Strumentazione Scientifca in Modena, Italy (cat. no. PC-035/91).	De Maddalena (1997, 2000c), Lawley (1881)
A few years before 1938	ITALY				Nonfatal attack on the boat of fisherman Maciotta.	Anonymous (1938)
1964	ITALY				Observed feeding on a young sperm whale. Captured.	Budker (1971), Fergusson (1996)
SLOVENIA						
August 24, 1938	Koper SLOVENIA	Ca. 500 (est.)			Encountered by fisherman Nicola Lubrano.	
Ca. 1940	Koper SLOVENIA				Attack on a boat.	M. Zuffa (pers. comm.)
October 22, 1963	Salvore, Golfo di Pirano SLOVENIA	Ca. 600 (492–547 TOT — est.)	Ca. 1100		Captured, killed with rifle shots. Stomach contained an estimated 200 kg dolphin. Photographic documentation exists.	Anonymous (1963), Bosnjak and Lipej (1992–1993), De Maddalena (2000b), De Maddalena et al. (2001), Lipej (1993–1994)

3 — Summary Tables

Date	Location	Length (cm)	Weight (kg)	Sex	Remarks	Sources
			CROATIA			
September 14, 1868	Jablanac CROATIA				Captured. Preserved in the Croatian Museum of Zagreb, Croatia.	Brusina (1888)
December 16, 1868	Sv. Gjuraj, Senj CROATIA	460			Captured. Preserved in the Croatian Museum of Zagreb, Croatia.	Brusina (1888)
April 16, 1872	Preluka CROATIA	490			Captured. Stomach contained the head and a leg of a man and a dolphin.	Brusina (1888), Hirtz (1932), Soldo and Jardas (2002)
May 12, 1872	Opuzen CROATIA	95			Captured.	Brusina (1888)
May 12, 1872	Konao, Mljet CROATIA	237			Captured.	Brusina (1888)
June 8, 1872	Preluka harbour CROATIA	131			Captured.	Brusina (1888)
June 16, 1872	Dugi Otok CROATIA	146			Captured.	Brusina (1888)
July 25, 1872	Cavtat CROATIA	260			Captured.	Brusina (1888)
August 8, 1872	Rab CROATIA	130			Captured.	Brusina (1888)
May 5, 1877	Ustrine, Cres CROATIA	460			Captured.	Brusina (1888)
May 8, 1877	Sv. Martin, Cres CROATIA	413			Captured.	Brusina (1888)
June 17, 1878	Osor, Cres CROATIA	371			Captured.	Ninni (1912), Perugia (1881)
August 9, 1878	Porec CROATIA				Captured?	Brusina (1888), Soldo and Jardas (2002)
May 21, 1879	Sv. Martin, Cres CROATIA	382			Captured.	Brusina (1888)
June 1879	Kvarner CROATIA				Captured.	Fergusson (1996), Graeffe (1886), Tortonese (1956)
July 23, 1879	Split CROATIA	402 or 445			Captured.	Ninni (1912), Perugia (1881)

Date	Location	Length (cm)	Weight (kg)	Sex	Remarks	Sources
October 21, 1879	Ustrine, Cres CROATIA	530			Captured.	Faber (1883), Fergusson (1996), Ninni (1912), Perugia (1881)
April 22, 1881	Rab CROATIA	380			Captured.	Brusina (1888)
October 16, 1881	Rab CROATIA	405			Captured.	Brusina (1888)
April 13, 1882	Sv. Martin, Cres CROATIA	529			Captured.	Brusina (1888)
June 13, 1883	Vrboska, Hvar CROATIA	300			Captured.	Brusina (1888)
September 26, 1883	Rab CROATIA	396			Captured.	Brusina (1888)
March 3, 1886	Korcula CROATIA	560			Captured.	Brusina (1888)
September 2, 1887	Krk CROATIA	470			Captured.	Brusina (1888)
July 1888	Sv. Juraj CROATIA	470			Captured. Stomach contained a woman and a lamb.	Hirtz (1932), Soldo and Jardas (2002)
April 26, 1891	Pag CROATIA				Captured. Preserved in the Natural History Museum of Zagreb, Croatia.	Langhoffer (1905), Soldo and Jardas (2002)
September 1892	Bakarac CROATIA	450			Captured?	Hirtz (1932), Soldo and Jardas (2002)
February 19, 1893	Zlarin CROATIA	165		M	Captured.	Katuric (1893), Soldo and Jardas (2002)
August 29, 1894	Bakar CROATIA	470		F	Captured. Preserved at the Prirodoslovni Muzej in Rijeka, Croatia (cat. no. PMR VP2).	De Maddalena (2006b, 2007), Kovačič (pers. comm.), Kovačič (1998), Soldo and Jardas (2002)
July 15, 1901	Dubrovnik CROATIA	520			Captured?	Kosic (1903), Soldo and Jardas (2002)
May 21, 1903	Senj CROATIA	Ca. 600	1200		Captured. Photographic documentation exists.	

3 — Summary Tables

Date	Location	Length (cm)	Weight (kg)	Sex	Remarks	Sources
May 29, 1906	Kvarner, Bakarac (?) CROATIA	522 TLn	1800?	F	Captured. Preserved as a taxidermied skin mount in the Museo di Storia Naturale in Trieste, Italy (without cat. no.).	Andrea dall'Asta (pers. comm.), De Maddalena (1999, 2000c) http://lokalpatrioti-rijeka.com
January 1908	Medola CROATIA				Fatal attack on Milena Scambelli fallen into the sea. Some doubt exists over the exact identity of the species.	Anonymous (1908), M. Zuffa (pers. comm.)
May 19, 1908	Stadival CROATIA	170			Captured by fishermen Simeone Armanini and Simeone Franceschini.	M. Zuffa (pers. comm.)
October 1909	Kraljevica CROATIA	550?			Captured in a tuna trap.	A. Mojetta (pers. comm.), Mojetta et al. (1997)
February 2, 1920	Dugi Otok, Kornati CROATIA	525	1300		Captured. Stomach contained a dolphin.	Soldo and Jardas (2002)
March 1926	Ugljan CROATIA	500	700		Captured?	Morovic (1950), Soldo and Jardas (2002)
August 1926	Lumbarda CROATIA	400	500		Captured. Stomach contained human remains.	Hirtz (1932), Morovic (1950), Soldo and Jardas (2002)
October 1926	Lumbarda CROATIA	600	1800		Captured in a net.	Morovic (1950, 1973), Soldo and Jardas (2002)
1927	Rovinj CROATIA	Ca. 600	1000		Captured. Stomach contained several inedible items. Photographic documentation exists.	De Maddalena (2000b)
1931	Rogoznica CROATIA	150		F	Captured?	Anonymous (1931), Soldo and Jardas (2002)

Date	Location	Length (cm)	Weight (kg)	Sex	Remarks	Sources
August 21, 1934	Susak CROATIA				Unprovoked fatal attack on bather Agnes Novak.	Giudici and Fino (1989)
August 23 and 30, 1934	Rijeka CROATIA	Ca. 600 (est.)	>700 (est.)		At least two sharks sighted by soldiers and fishermen.	Giudici and Fino (1989)
August 30, 1934	Rijeka CROATIA	600 (est.)			Unprovoked fatal attack on bather Zorica Princ.	Baldridge (1974), De Maddalena (2000b)
September 2, 1934	Kraljevica CROATIA	775 or >700	1100 or >2000		Captured in a tuna trap.	Giudici and Fino (1989), Morovic (1973), Soldo and Jardas (2002)
September 7, 1934	Moscenicka Draga CROATIA	600 or ca. 500	1000 or ca. 800		Captured in a tuna trap.	Anonymous (1934), Giudici and Fino (1989), Soldo and Jardas (2002)
September 7, 1934	Martinschizza CROATIA	>600 (est.)			Two sightings. Observed eating a small board of cork.	Giudici and Fino (1989)
July 20, 1935	Lukovo CROATIA	600	2500		Captured in a tuna trap.	Anonymous (1935), Morovic (1950), Soldo and Jardas (2002)
Summer 1946	Bakarac CROATIA				Captured. Stomach contained a 10 kg domestic pig.	Jolic (1988), Soldo and Jardas (2002)
Ca. 1950	Rovinj CROATIA				Sighted.	D. Abed-Navandi (pers. comm.)
August 1950	Primosten CROATIA	700–800 (est.)			Observed scavenging on the carcass of a calf.	Anonymous (1951), Soldo and Jardas (2002)
October 2, 1954	Pag CROATIA	550 (est.)	1500		Attack on a boat.	Anonymous (1954), Soldo and Jardas (2002)

3 — Summary Tables

Date	Location	Length (cm)	Weight (kg)	Sex	Remarks	Sources
December 10, 1955	Sveti Juraj, Senj CROATIA	460		F	Captured. Preserved as a taxidermied skin-mount at the Zemaljski Muzej Bosne i Hercegovine in Sarajevo, Bosnia and Herzegovina (without cat. no.).	De Maddalena (2006b, 2007), Đurović and Obratil (1984, 1989), D. Kotrošan (pers. comm.)
1956	Krk CROATIA	400			Captured?	Soldo and Jardas (2002)
September 24, 1961	Opatija CROATIA				Unprovoked fatal attack on bather Sabit Plana.	Anonymous (1961), Baldridge (1973), Giudici and Fino (1989), GSAF
August 16, 1966	Dalmatia CROATIA				Fatal attack.	A. Mojetta (pers. comm.)
1968	Rava CROATIA	Ca. 200 (est.)			Captured in a gill net. Photographic documentation exists.	Lončarić (pers. comm.), Lončarić (2009)
1970	Novigrad CROATIA				Unprovoked attack on spearfishing diver Jurincic.	A. Mojetta (pers. comm.)
1971	Ika CROATIA				Unprovoked fatal attack on bather Stanislaw Klepka.	Gilioli (1989)
August 17, 1972	Kornati CROATIA	600			Captured?	Morovic (1973), Soldo and Jardas (2002)
August 10, 1974	Omis CROATIA	Ca. 500 (est.)			Fatal attack on Rolf Schneider.	M. Zuffa (pers. comm.)
August 1993	Sibenik CROATIA	500		F	Captured in a net.	Fergusson (1996)
August 1993	Losinj CROATIA				Several sightings by fishermen.	Fergusson (1996), Mojetta et al. (1997)
Spring 1994	Brac CROATIA				Captured. Jaws preserved.	G. Della Rovere (pers. comm.)
September 4, 1998	Dubrovnik CROATIA				Sighted.	De Sabata et al. (1999)
August 2, 1998	Mljet CROATIA				Sighted.	De Sabata et al. (1999)

Date	Location	Length (cm)	Weight (kg)	Sex	Remarks	Sources
Spring 2000	Porec CROATIA	600 (est.)			Encountered by a fisherman.	D. Abed-Navandi (pers. comm.)
June 24, 2003	Jabuka CROATIA	570	Ca. 2500 (est.)	F	Captured in a tuna net by Pave Arkovic. Some teeth preserved. Photographic documentation exists.	ANSA, L. Lipej (pers. comm.), Soldo and Dulcic (2005)
October 6, 2008	Mala Smokova Bay, Vis CROATIA	Ca. 450–500 (est.)			Unprovoked non-fatal attack on spearfishing diver Damjan Pesek.	Branko Dragicevic (pers. comm.), Jakov Dulcic (pers. comm.)
MONTENEGRO						
June 1926	Herceg Novi MONTENEGRO	300			Captured. Stomach contained a woman's shoes and clothes.	Hirtz (1932), Morovic (1950), Soldo and Jardas (2002)
August 1955	Budva MONTENEGRO				Fatal attack.	Radovanovic (1965), Soldo and Jardas (2002)
August 5, 2011	Bar MONTENEGRO	217			Caught in a net. Photographic documentation exists.	http://sharkyear.com
GREECE						
A few years before 1878	Syros, Cyclades GREECE	Ca. 500			Observed eating some leather that was placed in the water near the shore close to a tannery. Captured. Said to be preserved at a museum in Athens.	Heldreich (1878)
Before 1943	Alexandroupolis GREECE				Captured?	Fergusson (1996), Konsuloff and Drenski (1943)
Before 1943	Kavallah GREECE	230			Captured?	Fergusson (1996), Konsuloff and Drenski (1943)
Before 1943	Thasos GREECE	180			Captured?	Fergusson (1996), Konsuloff and Drenski (1943)
September 22, 1948	Keratsini, Peiraus Port GREECE	Ca. 640 (est.)			Fatal attack on bather Dimitris Th. Parasakis.	Coppleson (1958), M. Bardanis (pers. comm.)

3 — Summary Tables

Date	Location	Length (cm)	Weight (kg)	Sex	Remarks	Sources
August 17, 1951	Mon Repo, Kerkira GREECE				Fatal attack on bather Vanda Pierri and non-fatal attack on bather George Athanasenas.	M. Bardanis (pers. comm.)
June 4, 1962	Aliki-Kato Axaia GREECE	480	2000		Captured. Photographic documentation exists.	M. Bardanis (pers. comm.)
June 1, 1963	Pithos o Micra Rock, Trikerion Island GREECE	Ca. 300 (est.)			Fatal attack on bather Helga Pogl, witnessed by Mr. Wilgen.	M. Bardanis (pers. comm.)
1969	GREECE				Fatal attack on a bather.	M. Bardanis (pers. comm.)
September 15, 1972	Makryalos, Thermaïkos Gulf GREECE	460	1300	M	Captured.	Economidis and Bauchot (1976)
Ca. 1985	Paliouri, Halkidiki GREECE	601–618 (est.)			Captured. Photographic documentation exists.	M. Bardanis (pers. comm.)
February 10, 1991	Kerkira GREECE	Ca. 600			Captured on a longline. Stomach contained an intact loggerhead sea turtle having a 60 cm diameter carapace. Jaws preserved. Photographic documentation exists.	M. Zuffa (pers. comm.)
August 1999	Othoni GREECE	Ca. 400–500			Encountered by Leonardo Leone De Castris from an inflatable boat	L. Leone De Castris (pers. comm.)
TURKEY						
February 1881	Beylerbeyi, Bosporus TURKEY	391			Stranded.	Fergusson (1996)
November 17, 1881	Bosporus TURKEY	470	1500	F	Captured.	Fergusson (1996)

Date	Location	Length (cm)	Weight (kg)	Sex	Remarks	Sources
1916	Salistra, Sea of Marmara TURKEY	Ca. 700			Captured in a tuna trap.	Devedjian (1945), Kabasakal (2003)
May 1920	Sedef Adas, Sea of Marmara TURKEY	465	Ca. 1200		Captured with a line intended for swordfish. Stomach contained an estimated 200 kg bluefin tuna, remains of a swordfish, some Atlantic bonito and a small stone.	Devedjian (1945), Kabasakal (2003)
Before 1926	Sea of Marmara TURKEY	Ca. 400			Captured with hand line intended for tuna. Stomach contained eight large Atlantic bonito.	Devedjian (1945), Kabasakal (2003)
February 20, 1926	Büyükada, Sea of Marmara TURKEY	450	>1500		Captured with hand line intended for tuna.	Kabasakal (2003)
1930	Bosporus TURKEY				Attack on a boat. Some doubt exists over the exact identity of the species.	Anonymous (1930), Xuereb (1998)
March 30, 1954	Tuzla, Sea of Marmara TURKEY	450	1500		Captured with hand line intended for tuna.	Kabasakal (2003)
April 15, 1956	Prince Island, Sea of Marmara TURKEY	618	Ca. 3000	F	Captured with hand line intended for tuna.	Kabasakal (2003)
February 1962	Bosporus TURKEY	>500	3750	F	Captured. Photographic documentation exists.	Fergusson (1996)
December 28, 1965	Dolmabahçe, Bosporus TURKEY	500	Ca. 4000	F	Captured with hand line intended for tuna.	Kabasakal (2003)
December 28, 1965	Maiden's Tower, Bosporus TURKEY	700	Ca. 3000	F	Captured with hand line intended for tuna.	Kabasakal (2003)
January 13, 1966	Kabataş, Bosporus TURKEY	Ca. 400	Ca. 2000		Captured with a a harpoon.	Kabasakal (2003)
January 13, 1966	Kabataş, Bosporus TURKEY	Ca. 400	Ca. 2000		Captured with a harpoon.	Kabasakal (2003)

3 — Summary Tables

Date	Location	Length (cm)	Weight (kg)	Sex	Remarks	Sources
1967	Büyükada, Sea of Marmara TURKEY				Captured. Photographic documentation exists.	Kabasakal (2008)
End of March 1968	Bosporus TURKEY	551 (est.)		F	Captured. Photographic documentation exists.	Kabasakal (2011)
May 1985	Kapidağ, Sea of Marmara TURKEY	Ca. 500 (est.)			Sighted.	Kabasakal (2003)
Ca. March 14, 1991	Foça, Aegean Sea TURKEY	Ca. 550	Ca. 3000		Captured. Stomach contained a tuna estimated to be 100 cm-long. Photographic documentation exists.	Anonymous (1991a), CLOMFOT (web site), Kabasakal (2008)
March 1996	Gökçeada Island, Aegean Sea TURKEY	550		F	Captured in a purse seine.	Kabasakal and Kabasakal (2004)
April 1998	Gökçeada Island, Aegean Sea TURKEY	Ca. 450 (est.)			Sighted by a fisherman.	Kabasakal and Kabasakal (2004)
May 1999	Golfo di Saros, Aegean Sea TURKEY	Ca. 500 (est.)			Encountered by a diver.	Kabasakal and Kabasakal (2004)
July 1, 2008	Edremit Bay, Edremit Bay TURKEY	125,5			Captured in a gill net. Stomach contained many embryonic teeth.	Kabasakal and Gedikoğlu (2008)
July 4, 2008	Edremit Bay, Edremit Bay TURKEY	145			Captured on a bottom longline. Stomach contained a few teeth and remains of a bony fish (probably the bait).	Kabasakal and Gedikoğlu (2008)
Before 1974	Prince Islands, Sea of Marmara TURKEY	Ca. 2000			Captured with hand line intended for tuna.	Güney (1974), Kabasakal (2003)
CYPRUS						
1993	Paphos CYPRUS	>500 (est.)		F?	Several sightings by divers.	Fisher et al. (1987)
ISRAEL						
Before 1971	Akko ISRAEL	200			Captured by hook and line. Jaws preserved.	Ben-Tuvia (1971)

Date	Location	Length (cm)	Weight (kg)	Sex	Remarks	Sources
			EGYPT			
Summer 1934	Agamy Beach, Al Iskandariyya (Alexandria) EGYPT	425	2500	F	Pregnant female carrying nine 61 cm-long embryos captured. Photographic documentation exists.	Compagno (1984), Ellis (1983), Lineaweaver III and Backus (1969), Mollet et al. (2000), Norman and Fraser (1937), Steel (1985), Tortonese (1956), Tricas and McCosker (1984)
			LIBYA			
June 12, 2002	Tripoli LIBYA	Ca. 500 (est.)		F?	Observed entering a tuna cage; two days later it was gone.	Galaz and De Maddalena (2004)
			MALTA			
Unknown date	MALTA				Captured. Three teeth preserved at the Zoologisk Museum in Copenhagen, Denmark (without cat. no.)	De Maddalena (2006b, 2007), J. Nielsen (pers. comm.)
April 25, 1890	Munxar reefs MALTA				Unprovoked fatal attack on the boat of fishermen Salvatore Bugeja, Agostino Bugeja, Carmelo Delia and Carmelo Abela. Salvatore Bugeja and Agostino Bugeja were never found. Some doubt exists over the exact identity of the species.	Gulia (1890), Xuereb (1998)
May 14, 1898	Mellieha MALTA	1429			Captured in a tuna trap. Some doubt exists over the exact identity of the species.	Anonymous (1898)
March 7, 1907	Marsaskala MALTA	Ca. 600 (est.)			Unprovoked fatal attack on a boat: two fishermen fell into the sea and were fatally attacked by the shark.	Mojetta et al. (1997)

3 — Summary Tables

Date	Location	Length (cm)	Weight (kg)	Sex	Remarks	Sources
September 1914	Wied iz-Zurrieq MALTA				Captured.	Xuereb (1998)
July 20, 1956	St. Thomas Bay MALTA	Ca. 500 (est.)			Unprovoked fatal attack on bather Jack Smedley, witnessed by Tony Grech and Alfred Xuereb.	Gianturco (1978), Kelleher (...), Mojetta et al. (1997), A. Xuereb (pers. comm.)
August 1956	Congreve Channel MALTA				Fatal attack on a boat of fishermen: two fishermen, Nazzareno Zammit and Emmanuel fell into the sea and Emmanuel was fatally attacked by the shark.	GSAF, M. Zuffa (pers. comm.)
May 1964	Filfla MALTA	Ca. 500 or 580	Ca. 2000	M	Captured with a line. Photographic documentation exists.	Fergusson (1996), Mojetta et al. (1997), A. Xuereb (pers. comm.)
March 1973	Filfla MALTA	530–550	1920	F	Captured in a net by Alfredo Cutajar. Photographic documentation exists.	Abela (1989), Fergusson (1996), Mizzi (1994), Mojetta et al. (1997), A. Xuereb (pers. comm.)
May 1979	Filfla MALTA	400		M	Captured. Photographic documentation exists.	Fergusson (1996), Mojetta et al. (1997), A. Xuereb (pers. comm.)
1982	Isola di Gozo MALTA	Ca. 500 (est.)			Sighted.	Mojetta et al. (1997)

Date	Location	Length (cm)	Weight (kg)	Sex	Remarks	Sources
April 17, 1987	Filfla MALTA	714 or 723 (668–681 TOT — est.)	2730 or 2880	F	Captured with a line by Alfredo Cutajar and Vince D'Amato. Stomach contained a whole 220 cm-long blue shark, a 250 cm dolphin that was consumed in two or three parts, a loggerhead sea turtle having a 60 cm diameter carapace, and a plastic bag containing garbage. Photographic documentation exists.	Abela (1989), Albertarelli (1990), Anonymous (1987), De Maddalena (1997), De Maddalena et al. (2001), Fergusson (1996, 1998b), Mollet et al. (1996)
Ca. April 1987	MALTA	Ca. 600 (est.)		M	Encountered by a fisherman.	Abela (1989)
August 18, 1993	MALTA	>500 (est.)		M	Sighted.	Fergusson (1996)
TUNISIA						
From 1953 to 1976	Golfe de Gabès TUNISIA				Captured?	Capapé et al. (1976), Fergusson (1996)
From 1953 to 1976	Kelibia TUNISIA				Captured?	Capapé et al. (1976), Fergusson (1996)
From 1953 to 1976	La Galite TUNISIA				Some sharks captured.	Capapé et al. (1976), Fergusson (1996)
20 May (ca.) 1953	Sidi Daoud TUNISIA	600?	1500		Captured in a tuna trap.	Postel (1958)
Ca. May 20, 1954	Sidi Daoud TUNISIA				Two sharks captured in a tuna trap.	Postel (1958)

3 — Summary Tables

Date	Location	Length (cm)	Weight (kg)	Sex	Remarks	Sources
May 16, 1956	Sidi Daoud TUNISIA	520	1800	F	Captured in a tuna trap. Stomach contained skeletons of two common dolphins (an adult and probably an embryo), parts of a green sea turtle, and a small mako shark (most likely a shortfin mako) that measured a little over 1 m. Photographic documentation exists.	Postel (1958)
May 22, 1956	Sidi Daoud TUNISIA	410	1300	M	Captured in a tuna trap.	Postel (1958)
Ca. 1970	Zembra Island TUNISIA				Some sharks captured.	Capapé et al. (1976), Fergusson (1996)
Ca. 1970	Bizerte TUNISIA				Some sharks captured.	Capapé et al. (1976), Fergusson (1996)
Between 1970 and 1980	Ras Fartas TUNISIA	523		M	Captured.	Bradaï et al. (2002)
March 21, 1972	Zarzis TUNISIA	400			Captured.	Quignard and Capapé (1972)
March 21, 1972	Zarzis TUNISIA	400			Captured.	Quignard and Capapé (1972)
April 21, 1972	Zarzis TUNISIA	185			Captured.	Quignard and Capapé (1972)
May 1979	Northwest of Cap Bon TUNISIA	220		F	Captured?	Fergusson (1996)
June 1979	Northwest of Cap Bon TUNISIA	>500	(ca. 520)	F	Captured. Stomach contained a dolphin.	Fergusson (1996)
August 1981	Northwest of Cap Bon TUNISIA		185	M	Captured.	Fergusson (1996)
June 1983	Northwest of Cap Bon TUNISIA		175	F	Captured.	Fergusson (1996)
April 1985 Northwest of Cap Bon	TUNISIA		190	F	Captured.	Fergusson (1996)

Mediterranean Great White Sharks

Date	Location	Length (cm)	Weight (kg)	Sex	Remarks	Sources
July 1988	Golfe de Gabès TUNISIA	220	165	M & M	Two sharks captured.	Bradaï et al. (2002)
February 4, 1989	Ghar El Melh TUNISIA	670 or 570	2032	F	Captured.	Anonymous (1989a), Giudici and Fino (1989)
September (?) 1992	Sidi Daoud TUNISIA	>500		F	Pregnant female carrying two full term embryos captured.	Fergusson (1996)
December 14, 1992	Golfe de Gabès TUNISIA	435		M	Captured in a purse seine.	Bradaï et al. (2002)
February 18, 1993	Golfe de Gabès TUNISIA	425		M	Captured in a purse seine.	Bradaï et al. (2002)
August 1993	Skerki Bank TUNISIA	Ca. 600 (est.)			Encountered by fishermen. Photographic documentation exists.	Fergusson (1998c)
February 9, 1997	Golfe de Gabès TUNISIA	425		F	Captured in a purse seine.	Bradaï et al. (2002)
February 26, 1997	Golfe de Gabès TUNISIA	179		M	Captured in a trawl.	Bradaï et al. (2002)
December 25, 1997	Golfe de Gabès TUNISIA	170		M	Captured in a trawl.	Bradaï et al. (2002)
May 26, 1998	Sidi Daoud TUNISIA	Ca. 250	Ca. 225	F	Captured in a tuna trap. Photographic documentation exists.	Fergusson (1998a)
April 23, 2001	Sidi Daoud TUNISIA	490	2200	M	Captured in a tuna trap. Stomach contained the vertebral column of a tuna. A tooth from this shark has been preserved by Philippe Laulhe.	P. Laulhe (pers. comm.), W. Maamouri (pers. comm.)
April 27, 2001	Sidi Daoud TUNISIA	520	2250	F	Captured in a tuna trap. Stomach contained a large sea turtle. A tooth from this shark has been preserved by Philippe Laulhe. Photographic documentation exists.	De Maddalena and Reckel (2003), P. Laulhe (pers. comm.), W. Maamouri (pers. comm.)

3 — Summary Tables

Date	Location	Length (cm)	Weight (kg)	Sex	Remarks	Sources
February 26, 2004	Golfe de Gabès TUNISIA	587		F	Pregnant female carrying four 132–135 cm-long embryos captured in a a purse seine. Photographic documentation exists. One of the embryos was preserved at the Institut national des sciences et technologies de la mer, Centre de Sfax, Tunisia (cat. no. INSTM/LAM 01).	Saidi et al. (2005)
April 20, 2006	Aras Dizra TUNISIA		Ca. 350 (gutted)	F	Captured in a net by Mohamed Chelbaïa. Tissue sample preserved at the Muséum national d'Histoire naturelle in Paris, France (tissu: BPS-0539).	Iglésias (2001)
2008	Chebba TUNISIA	420 (est.)	Ca. 1500	M	Captured. Film documentation exists.	
October 18, 2009	Mahdia TUNISIA	480	1426	F	Observed entering a tuna cage and attacking the tuna (October 17, 2009). The divers attempted to release the shark, but one of them hit the animal near the eye with a knife and the next day (October 18, 2009) it was found dead. Stomach contained remains of tuna and albatrosses. Photographic and film documentation exists.	M. Dahmouni (pers. comm.)

Date	Location	Length (cm)	Weight (kg)	Sex	Remarks	Sources
			ALGERIA			
Before 1865	ALGERIA		223	M	Captured. This is probably the white shark preserved as a taxidermied skin-mount at the Museum National d'Histoire Naturelle in Paris, France (cat. no. MNHN a-9695).	De Maddalena (2006b, 2007), Dieuzeide and Novella (1953), Dumeril (1865), B. Séret (pers. comm.), M. Zuffa (pers. comm.)
1882	Alger ALGERIA				Captured. Jaws preserved at the Zoologisk Museum in Copenhagen, Denmark (without cat. no.)	De Maddalena (2006b, 2007), J. Nielsen (pers. comm.)
Before 1999	ALGERIA	150			Captured.	M. Hamdine (pers. comm.)
Before 2006	ALGERIA				Captured. Jaws preserved at the Museum National d'Histoire Naturelle in Paris, France (cat. no. MNHN a-9922).	De Maddalena (2006b, 2007), B. Séret (pers. comm.), M. Zuffa (pers. comm.)
2008 (?)	Ghazaouet ALGERIA			F	Captured in a trawl. Film documentation exists.	
			MOROCCO			
Before 1926	Golfe de Chafarinas MOROCCO				Captured?	Fergusson (1996)
Early March 1980	Tetouan MOROCCO	555			Captured. Photographic documentation exists.	M. Zuffa (pers. comm.)

3 — Summary Tables

Date	Location	Length (cm)	Weight (kg)	Sex	Remarks	Sources
Unknown location						
From April 1872 to July 1882	Northeastern Adriatic	From 146 to 530			21 sharks captured. Surely some of these sharks, or possibly all, have already been accounted for among the Italian and Croatian records from the Adriatic Sea	Marchesetti (1884)
June 17, 1879	Adriatic				Captured.	Ninni (1912), Perugia (1881)
June 17, 1879	Adriatic				Captured.	Ninni (1912), Perugia (1881)
1901	Eastern Adriatic	500			Captured?	Barac (1901), Soldo and Jardas (2002)
May 1947	Eastern Adriatic	300	300		Captured?	Anonymous (1947), Soldo and Jardas (2002)
1969	Eastern Adriatic				Captured?	Morovic (1969)
Unknown date	Adriatic				Captured. Jaws preserved at the Museo di Storia Naturale in Trieste, Italy (without cat. no.).	De Maddalena (2000c)
Before September 1891	Adriatic	1005	4000		Captured.	Anonymous (1891), Ellis and McCosker (1991)
Before 1919	Adriatic	438		F	Captured. Stomach contained the boots of a seaman from the Austro-Hungarian navy. Preserved as a taxidermied skin-mount at the Naturhistorische Museum in Vienna, Austria (cat. no. NMW-95054).	De Maddalena (2006b, 2007), H. Wellendorf (pers. comm.)

Date	Location	Length (cm)	Weight (kg)	Sex	Remarks	Sources
Before 1969	Adriatic				Captured. Stomach contained a raincoat, two or three coats and a car licence plate.	De Maddalena (2000b), Gianturco (1978), Lineaweaver III and Backus (1969)
16th Century (?)	Mediterranean	>500			Captured. Stomach contained a seaman and two tunas.	Brunnich (1768), Steel (1985)
1984	Mediterranean				Captured on a longline.	Rey et al. (1986)
1998–2001	Eastern Mediterranean	429			Captured on a longline.	Megalofonou et al. (2005)
Before 1640–1660	Mediterranean?	450–640 (est.)			Captured. Jaws preserved at the Biblioteca Ambrosiana in Milan, Italy (without cat. no.) (originally preserved in the ancient Settala Museum in Milan).	Anonymous (1640–1660), De Maddalena (2000c, 2006a), Terzago and Scarabelli (1677)
Second half of the 18th Century	Mediterranean?				Captured. Lower jaw preserved at the Musei Civici in Reggio-Emilia, Italy (cat. no. 14–4527).	De Maddalena (2000c)
1758	Mediterranean	600			Nonfatal attack on a sailor who fell overboard from a frigate. Captured with a harpoon. Exhibited throughout Europe.	GSAF
19th Century	Mediterranean?				Captured. Upper jaw preserved at the Museo di Storia Naturale in Venezia, Italy (cat. no. 4860 Ist.V.S.L.A. 133).	De Maddalena (2000c), Mizzan (1994), Trois (1900)

3—Summary Tables

Date	Location	Length (cm)	Weight (kg)	Sex	Remarks	Sources
1868 or between 1872 and 1878	Mediterranean?				Captured. Jaws preserved at the Museo Zoologico in Napoli, Italy (cat. no. Z6431).	De Maddalena (2006b, 2007), N. Maio (pers. comm.), Maio et al. (2005)
1868	Mediterranean?				Captured. Brain preserved at the Museo Zoologico in Napoli, Italy (cat. no. Z1113).	Maio et al. (2005)
1876	Mediterranean?				Captured. Eye preserved at the Museo Zoologico in Napoli, Italy (cat. no. Z1194).	Maio et al. (2005)
1876 or 1878	Mediterranean?				Captured. Heart preserved at the Museo Zoologico in Napoli, Italy (cat. no. Z1244).	Maio et al. (2005)
Second half of May 2000	Mediterranean?	Ca. 500			Captured. Some doubt exists over the exact identity of the species.	De Maddalena and Piscitelli (2001)
Unknown date	Mediterranean				Captured. Jaws preserved at the Centre de Conservation et d'Etude des Collection, Musée des Confluences in Lyon, France.	D. Berthet (pers. comm.)
Before 1870	Mediterranean?				Captured. Jaws preserved at the Museo di Anatomia Comparata in Rome, Italy (cat. no. 111–167) (originally preserved in the ancient Pontificium Romanum Archigymnasium in Rome, Italy).	E. Capanna (pers. comm.), De Maddalena (2000c, 2006b, 2007)

Mediterranean Great White Sharks

Date	Location	Length (cm)	Weight (kg)	Sex	Remarks	Sources
Before 1870	Mediterranean?				Captured. Brain preserved at the Museo di Anatomia Comparata in Rome, Italy (cat. no. 111–208) (originally preserved in the ancient Pontificium Romanum Archigymnasium in Rome, Italy).	E. Capanna (pers. comm.), De Maddalena (2000c, 2006b, 2007)
Before 1707	Mediterranean?				Captured. Cranium preserved at the Museo di Anatomia Comparata in Rome, Italy (without cat. no.) (originally preserved in the ancient Museum Kircherianum).	Bonanni (1707), E. Capanna (pers. comm.), De Maddalena (2000c, 2006b, 2007)
Unknown date	Mediterranean?				Captured. Jaws preserved at the Museo di Storia Naturale in Genova, Italy (without cat. no.).	De Maddalena (1997, 2000c)
Unknown date	Mediterranean?				Captured. Cranium preeserved at the Museo di Anatomia Comparata in Torino but belongs to the Museo Regionale di Scienze Naturali in Torino, Italy (cat. no. 612).	De Maddalena (2000c, 2002)
Unknown date	Mediterranean?				Captured. Part of an upper jaw preserved at the Museo di Storia Naturale e della Strumentazione Scientifica in Modena, Italy (cat. no. -049/91).	De Maddalena (2000c)

3 — Summary Tables

Date	Location	Length (cm)	Weight (kg)	Sex	Remarks	Sources
Before 20th Century?	Mediterranean?	Ca. 200 (est.)			Captured. Cranium preserved at the Museo di Storia Naturale e del Territorio in Calci, Italy (without cat. no.).	De Maddalena (2000c)
Unknown date	Mediterranean?	Ca. 200 (est.)			Captured. Jaws preserved at the Museo di Storia Naturale in Livorno, Italy (without cat. no.).	De Maddalena (1997, 2000c)
Unknown date	Mediterranean?				Captured. Jaws preserved at the Museo Zoologico in Florence, Italy (cat. no. 6361).	De Maddalena (2000c), Vanni (1992)
Unknown date	Mediterranean?				Captured. Jaws preserved at the Istituto di Zoologia in Genova, Italy (without cat. no.).	De Maddalena (2000c)
Unknown date	Mediterranean?				Captured. Jaws preserved at the Istituto di Zoologia in Genova, Italy (without cat. no.).	De Maddalena (2000c)
Unknown date	Mediterranean?				Captured. Jaws preserved at the Museo di Storia Naturale in Ferrara, Italy.	De Maddalena (2000c)
Unknown date	Mediterranean?	500		F	Captured. Preserved as a taxidermied skin-mount at the Muséum National d'Histoire Naturelle in Paris, France (cat. no. MNHNa-4669).	De Maddalena (2006b, 2007), B. Séret (pers. comm.)

Date	Location	Length (cm)	Weight (kg)	Sex	Remarks	Sources
Unknown date	Mediterranean?				Captured. Jaws preserved at the Muséum National d'Histoire Naturelle in Paris, France (cat. no. MNHNab-0002).	De Maddalena (2006b, 2007), B. Séret (pers. comm.)
Unknown date	Mediterranean?				Captured. Jaws preserved at the Muséum National d'Histoire Naturelle in Paris, France (cat. no. MNHNab-0003).	De Maddalena (2006b, 2007), B. Séret (pers. comm.)
Unknown date	Mediterranean?				Captured. Jaws preserved at the Muséum National d'Histoire Naturelle in Paris, France (cat. no. MNHNab-0004).	De Maddalena (2006b, 2007), B. Séret (pers. comm.)
Unknown date	Mediterranean?				Captured. Jaws preserved at the Muséum National d'Histoire Naturelle in Paris, France (cat. no. MNHNab-0143).	De Maddalena (2006b, 2007), B. Séret (pers. comm.)
Unknown date	Mediterranean?		383		Captured. Preserved at the Muséum National d'Histoire Naturelle in Paris, France.	Dumeril (1865)
Unknown date	Mediterranean?				Captured. Jaws preserved at the Muséum d'Histoire Naturelle in Grenoble, France (cat. no. MHNGr.OS.47).	P. Candegabe (pers. comm.), De Maddalena (2006b, 2007)
Unknown date	Mediterranean?				Captured. Jaws preserved at the Muséum d'Histoire Naturelle de Nîmes, France (cat. no. 20).	De Maddalena (2006b, 2007), G. Gory (pers. comm.)

3 — Summary Tables

Date	Location	Length (cm)	Weight (kg)	Sex	Remarks	Sources
Unknown date	Mediterranean?				Captured. 25 teeth preserved at the Musée Océanographique de Monaco, Monaco.	M. Bruni (pers. comm.), De Maddalena (2006b, 2007)
Unknown date	Mediterranean?				Captured. Head preserved at the Muséum Requien in Avignon, France (cat. no. MR 1996–204)	E. Crégut (pers. comm.), De Maddalena and Zuffa (2008)
Unknown date	Mediterranean?				Captured. Lower jaw preserved at the Muséum d'Histoire Naturelle in Nice, France (cat. no. OSTEO-130).	De Maddalena and Zuffa (2008), O. Gerriet (pers. comm.)
Unknown date	Mediterranean?				Captured. Jaws preserved at the Muséum d'Histoire Naturelle in Nice, France (cat. no. OSTEO-220).	De Maddalena and Zuffa (2008), O. Gerriet (pers. comm.)
Unknown date	Mediterranean?				Captured. Jaws preserved at the Muséum d'Histoire Naturelle in Valence, France (cat. no. 2006.00.713).	De Maddalena and Zuffa (2008), P. Soleil (pers. comm.)
Before 1863	Mediterranean?				Captured. Jaws preserved at the Musei Civici in Reggio-Emilia, Italy (cat. no. 11-4?4).	De Maddalena (2000c)
Before 1881	Mediterranean	420			Captured. Preserved as a taxidermied skin-mount at the Museo di Storia Naturale in Genova, Italy.	De Maddalena (2000c), Moreau (1881)
Before 1881	Mediterranean?	548	1814		Captured.	Moreau (1881)

Date	Location	Length (cm)	Weight (kg)	Sex	Remarks	Sources
Before May 1, 1900	Mediterranean				Captured. Chondrocranium, jaws and vertebrae preserved at the Koninklijk Belgisch Instituut voor Natuurwetenschappen in Brussels, Belgium (cat. no. 1385 C).	De Maddalena (2006b, 2007), G. Lenglet (pers. comm.)
Before September 18, 1902	Mediterranean?				Captured. Jaws preserved at the British Museum of Natural History in London, United Kingdom (cat. no. BM(NH) 1905.12.2.2).	O. Crimmen (pers. comm.), De Maddalena (2006b, 2007)
Before 1909	Mediterranean?	223			Captured. It was preserved as a taxidermied skin-mount at the Museo di Storia Naturale in Genova, Italy.	Condorelli and Perrando (1909), De Maddalena (2000c)
Before 1909	Mediterranean?				Captured. Jaws preserved at the Museo di Storia Naturale in Palermo, Italy (likely one of the five jaw sets with cat. nos. An-108, An-115, An-128, An-145, and An-80). Doubtful case.	Condorelli and Perrando (1909), De Maddalena (2006b, 2007)

4

Analysis of the Presence of Great White Sharks in the Mediterranean Sea

Common Names Used in the Mediterranean Area

The common names given to the great white shark in the Mediterranean Countries follow: peshkagen njeringrenes (Albanian), kalb (Arabic), bijela ajkula (Bosnian), tauró blanc (Catalan), velika bijela psina, pas ljudozder (Croatian), wahsh (Egyptian), great white shark, white shark, white pointer, blue pointer, white death (English), grand requin blanc (French), weißer hai (German), lefkos karkarias (Greek), qarha levana (Hebrew), squalo bianco, grande squalo bianco, pescecane (Italian), kelb abjad (Maltese), kalb (Moroccan), velika bijela ajkula (Serbian), beli morski volk (Slovenian), tiburón blanco, jaquetón blanco (Spanish), kelb el b'har (Tunisian), canavar köpekbaligi (Turkish) (Barrull and Mate, 2002; De Maddalena and Baensch, 2005; Ellis and McCosker, 1991; Adem Hamzic, pers. comm.; Zoran Kljajic, pers. comm.; Lipej et al., 2004).

Similar Species Present in the Study Area

The identity of the great white shark is unmistakable. Nevertheless, there are some other sharks that show strong similarities to this species because they are closely related. These species belong to the genera Isurus or Lamna, and belong to the family Lamnidae, which includes the shortfin mako *Isurus oxyrinchus*, longfin mako *Isurus paucus*, porbeagle *Lamna nasus* and the salmon shark *Lamna ditropis*. Three of these species have been recorded in the Mediterranean Sea. The shortfin mako is fairly common, the porbeagle is scarce, the longfin mako is very rare and the salmon shark does not occur in these waters.

All lamnid sharks attain a large size and have a spindle-shaped body, conical snout, dark circular eyes, prominent lower teeth, caudal keels, and a lunate caudal fin. The similarities among the species of the family Lamnidae are not so pronounced as to create problems in identification, except when a novice observer examines a very young white shark. The sharks of the genus Isurus (*Isurus oxyrinchus* and *Isurus paucus*) have a slender body and a narrow snout. The teeth are strongly pointed, curved backward and narrow with unserrated cutting margins. The coloration can be bright blue with a strong metallic appearance (*Isurus oxyrinchus*) or dull grey-blue to black (*Isurus paucus*). The sharks of the genus *Lamna* (*Lamna nasus* and *Lamna ditropis*) have two pairs of caudal keels that are also present in *Isurus paucus* but not those of *Isurus oxyrinchus*, which have a single pair. Caudal keels are horizontal extensions in the area of the caudal peduncle, or base of the tail, that are thought to give horizontal stability to these fast-swimming sharks. The teeth of the genus *Lamna* sharks are small, lack serrations and have two small lateral cusplets. The coloration of the body can be bluish-grey to black and, in *Lamna nasus*, a free rear tip of the first dorsal fin has a conspicuous white patch (De Maddalena, 2007; De Maddalena, et al., 2005; De Maddalena and Baensch, 2005). Although lamnid sharks can reach a large size, the white shark dwarfs the others. The porbeagle reaches a maximum length of 360 cm. The longfin mako reaches a maximum length of 417 cm. Shortfin makos reach a maximum length of at least 577 cm. White sharks can reach a maximum length of at least 668 cm.

In the Mediterranean area, the general public cannot distinguish one species of the family Lamnidae from another. This is primarily due to the fact that the great white shark has never been abundant in the area and, with some exceptions, close encounters with humans are rare. Also, in some local dialects, a single common name was widely used for any species of the family Lamnidae for a very long time. So words like *pescecane* in Italian and *haifisch* in German were used for centuries to refer to all species of sharks of the family Lamnidae. In general, it is evident that popular knowledge of sharks in the Mediterranean area is very limited. Even experienced fishermen often mistake young white sharks for similarly sized porbeagles and shortfin makos. This is the main reason why most captures of juvenile white sharks go unnoticed.

Maximum Size

The 589 cm TOT female great white shark captured off Maguelone and landed in Sète, France, is the largest preserved specimen available today and is housed in the Museum of Zoology in Lausanne, Switzerland (without a catalogue number). This specimen is preserved as a cast that was constructed directly from the original whole specimen. The shark was accurately measured at the time of capture and is still measurable today from the cast, making it the largest specimen ever reported with a size that cannot be disputed (De Maddalena et al., 2003). The analysis of the sighting and capture data for great white sharks from the Mediterranean Sea has shown that, of the 593 records of great white sharks, 81 sharks were reported to be measured or estimated as being larger than the 589 cm TOT, the size of the Maguelone specimen. In most cases, the reliability of the reported size is impossible to verify and cannot be accepted or refuted. For instance, there was a huge white shark caught in 1947 off Scopello, Sicily, Italy, with a reported weight of 3590 kg, which is consistent with a shark over 700 cm TL (according to Mollet and Cailliet, 1996). The problem with the record is

This estimated 668–681 cm female white shark was caught on April 17, 1987, near Filfla, Malta (photograph by John Gullaumier, courtesy In-Nazzjon, Malta paper).

that it is not known how the shark was weighed and the reported length of 900–1000 cm was not measured but merely estimated. The photographic documentation does not provide any reliable estimation of the shark size. Therefore, it is impossible to determine the real size of this large shark.

Nevertheless, the size reported for some of these huge specimens should be considered reliable, based on the available photographic evidence. The cases with the largest sharks that are deemed reliable are reported as follows in the order of decreasing size: A female shark caught at the Estaque, a quarter of Marseille, on October 15, 1925 (De Maddalena and Zuffa, 2008; Philippe Summonti, pers. comm.), was estimated to be 667–687 cm TOT, a female caught near Filfla, Malta, on April 17, 1987, was estimated to be 668–681 cm TOT (De Maddalena et al., 2001), a female shark caught off Ganzirri, Italy, on June 19, 1961, was estimated at 666 cm TOT (De Maddalena et al., 2001); a female shark caught off Majorca, Spain, in March 1969 was estimated at 620 cm TL (Morey et al., 2003) and a female shark caught off Majorca, Spain, on February 5, 1976, was esti-

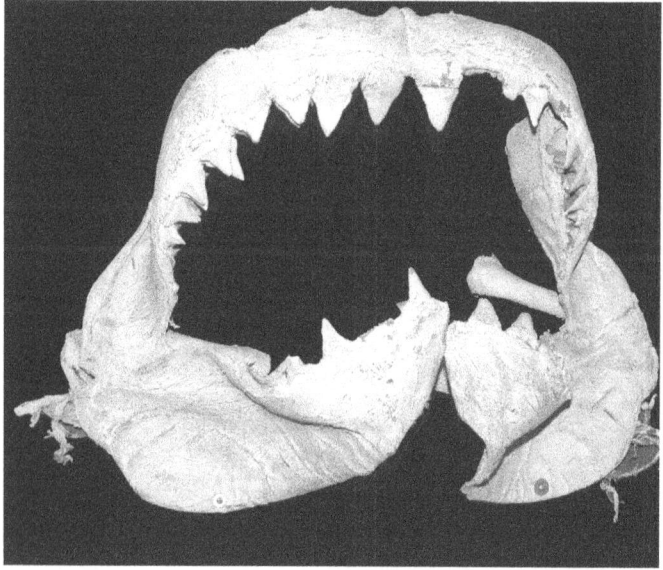

Top: This 620–640 cm white shark was caught in May 1974 near Isola Formica, Egadi, Italy (courtesy Nitto Mineo). *Bottom:* This set of jaws from a female white shark in excess of 600 cm that was either captured or stranded in early February of 1839 in Civitanova Marche, Italy, is preserved at the Museo di Anatomia Comparata in Rome, Italy (cat. no. 111-95) (photograph by Ernesto Capanna, courtesy Museo di Anatomia Comparata di Roma, Italy).

mated to be 610 cm TL (Morey et al., 2003); a shark caught off Paliouri, Halkidiki, Greece, around 1985 (Manolis Bardanis, pers. comm.) was estimated to be 601–618 cm TOT. Today, from Mediterranean specimens, there is solid evidence that great white sharks attain at least 668 cm TOT (De Maddalena et al., 2001).

Average Size

The average size of the sharks recorded in the great white shark data bank for the Mediterranean cannot be considered reliable. Reports of large sharks are more numerous than those for smaller sharks. This happens for different reasons. As stated previously, fishermen often mistake juvenile white sharks for similarly sized porbeagles and shortfin makos. Another reason is that the capture or sighting of a large shark causes a greater stir than that for a small specimen. A 200 cm white shark goes unreported because its size is not exceptional, falling well in the size range of many sharks commonly caught in the Mediterranean. However, 400 cm-long sharks are usually reported, regardless of their species. Mediterranean sharks that can attain this length are great white sharks, shortfin mako shark *Isurus oxyrinchus*, basking sharks *Cetorhinus maximus*, and bluntnose sixgill sharks *Hexanchus griseus*.

Color

So far, there appears to be no difference in coloration of the sharks observed in Mediterranean specimens compared to sharks inhabiting different geographical areas. Only a single case of albinism in *Carcharodon carcharias*, a specimen completely white with red eyes, has been recorded and it was in South African waters (Smale and Heemstra, 1997). In the Mediterranean, no confirmed case of albinism has been reported. However, a shark encountered several times in September 1986 off Rimini was described by different witnesses as having a white coloration. Unfortunately, it has not been possible to find any photographic documentation of this shark.

Reproduction

In great white sharks, males are mature between 350 cm and 410 cm TL, which corresponds to between 9 and 10 years of age. Females mature from 400 cm to 500 cm TL, which corresponds to between 12 and 14 years of age (Compagno, 2001). No white shark mating has been observed in the Mediterranean. During mating, the male white shark begins courtship by approaching the female and grabbing her with his mouth, causing superficial wounds called love bites or mating scars (Francis, 1996). Scars that

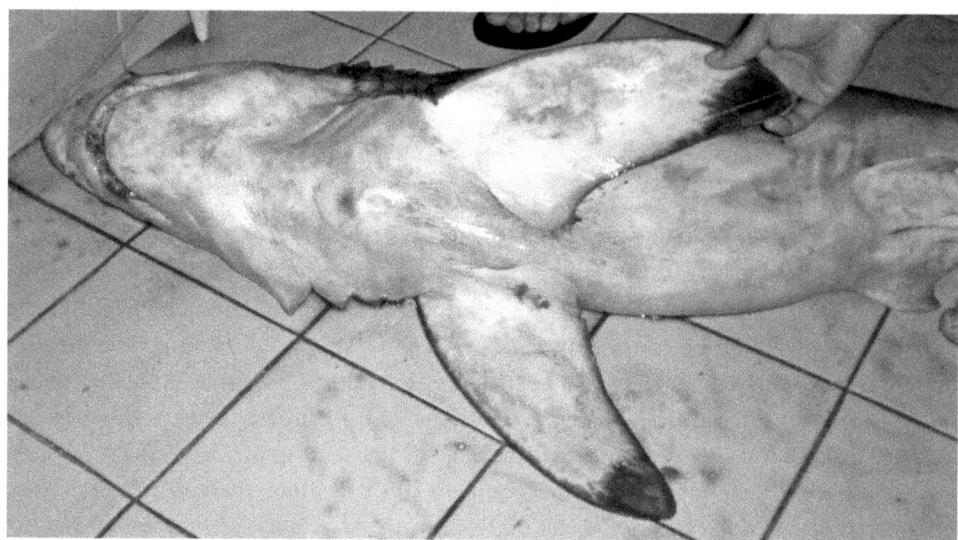

This approximately 130 cm long newborn white shark was caught in early July 2008 in the Edremit Bay, Turkey (photograph by Hakan Kabasakal).

were probably the result of love bites were observed on some females encountered or captured in the Mediterranean. These scars were located on the head, trunk, and pectoral fins.

Records of pregnant female white sharks are incredibly rare, both in the Mediterranean and in other areas, which is a mystery. It is clear that the large size of pregnant female white sharks make them hard to capture. It is possible that during gestation, the females abandon the rest of the population, going to remote areas far offshore or into deep waters. It is also possible that they stop feeding temporarily when they give birth to their pups, so that the probability of contact with humans is remote (De Maddalena, 2002). These hypotheses are speculation and are insufficient to explain this mystery. However, the scarcity of pregnant female white sharks suggests that they have very low fecundity.

Only four pregnant females were recorded in the Mediterranean, of which one is doubtful. Of these four females, three were recorded in the Canale di Sicilia (Channel of Sicily) with two off Tunisia and one off Sicily. The fourth was recorded off Egypt. The earliest report was off Egypt in the summer of 1934, when a 425 cm pregnant female carrying nine embryos was caught off Al Iskandariyya (Alexandria), Egypt. The embryos were reported measuring 61 cm and weighing 49 kg (108 lbs) (Norman and Fraser, 1937). The weight is obviously wrong. Tricas and McCosker (1984) hypothesized that the original source made a mistake and that 49 kg was the total weight of the nine embryos, so the average weight of each embryo was 5.4 kg. Another possibility is that the correct weight for each embryo was 4.9 kg, which was written as 49 kg by mistake. On an unknown date prior to 1981, a pregnant female carrying six 40 cm-long embryos was caught off Sciacca, Italy (Di Milia, 1981). However, there is some doubt over the exact identity of the species. In 1992, perhaps in September, a pregnant female carrying two full-term embryos and estimated at over 500 cm was caught northwest of Cap Bon, Tunisia. Neither the female nor the embryos were examined or measured by researchers (Fergusson, 1996). On February 26, 2004, a 587 cm pregnant female carrying four embryos was caught in the Golfe de Gabès, Tunisia. The embryos were three females and one male (one male and one female in the right uterus and two females in the left uterus). The abdomens of the embryos were considerably distended by the yolk mass. The embryos ranged in length from 132 cm to 135 cm and in weight from 27.65 kg to 31.50 kg. One of the embryos was preserved at the Institut national des sciences et technologies de la mer, Centre de Sfax (cat. no. INSTM/LAM 01) (Saidi et al., 2005).

Francis (1996) suggested that parturition may occur in the spring or summer in temperate locations of both hemispheres. Data from the Mediterranean confirmed this hypothesis. Great white shark pups range in size from 81.3 cm to 151 cm TL at birth (Ralph S. Collier, pers. comm.; Francis, 1996; Uchida et al., 1996). The number of white sharks recorded measuring 150 cm or less from the Mediterranean, which can be considered newborn, was 20. After birth, the pups show an umbilical scar. Thus, a newborn pup is identified by the presence of an unhealed umbilical scar (Kabasakal, 2008). The presence of an umbilical scar was reported for two of the 20 newborn sharks from the study area (Kabasakal and Gedikoğlu, 2008). For 13 of these newborn sharks, the capture date was recorded, and 12 of the 13 sharks were captured in the late spring and summer months, from May to August. The capture months are as follows: two sharks in May, one of which is doubtful, three in June, three in July and four in August. The exception to this time period is one record in November. From the data, it is evident that in the Mediterranean, female great white sharks give birth to their pups from May to August.

Of the 20 newborn pups, 9 were recorded in the Canale di Sicilia (six off Sicily, two off Lampedusa and one off Algeria), six in the northeastern Adriatic (five off Croatia and one in the northeastern Adriatic at an unspecified location), and the others were distributed among Alboran Sea (1), Ligurian Sea (1), Tyrrhenian Sea (1) and Aegean Sea (2). The 9 newborn sharks recorded in the Canale di Sicilia demonstrate that, in the Mediterranean, the Canale di Sicilia is the primary parturition ground for female white sharks. This supports the hypothesis previously proposed by Cigala Fulgosi (1990). The newborn sharks recorded in the Northeastern Adriatic suggest that in the past, up to the early 20th century, great white sharks pupped even in Croatian waters, but this does not seem to occur presently in those waters.

Litter sizes of the great white shark range from 2 to 14 and perhaps as high as 17 (Cliff et al., 2000). The few litters recorded from the Mediterranean ranged from 2 to 9 pups, which is consistent with the overall observed range. Two newborn pups measuring 80 cm and 95 cm have been reported in the Mediterranean (Brusina, 1888; Angelo Mojetta, pers. comm.; Mojetta et al., 1997) and that size is close to the minimum newborn white shark size reported elsewhere. Unfortunately, both records of these cases were poorly documented. Besides a paturition ground, the Canale di Sicilia seems to also represent a nursery area for newborn and juvenile white sharks. In fact, of the 19 juvenile sharks ranging in size from 151 to 200 cm in length, 10 were recorded in the Canale di Sicilia, while the others were distributed among the Tyrrhenian Sea (4), Croatia (3), Greece (1) and Israel (1).

With the exception of this nursery area, the data collected does not indicate any segregation by size in the study area. The gender of the shark has been reported for 138 specimens, of which 47 were males and 91 were females. The corresponding sex ratio was 1 male to 1.94 females. The sex ratio was particularly unbalanced in two regions. In Croatian waters, the sex ratio was 1 male to 6 females and in the waters of the Isole Baleari, the sex ratio was 1 male to 9 females.

Distribution

The records from the great white shark data bank have shown that the great white shark is present throughout the entire Mediterranean Sea. Today, depending on the area, the presence is considered rare or very rare (see page 214, "Abundance"). The distribution of white sharks has been primarily based on the boundaries of the nations. Having an extensive coastline, Italy is the nation with the highest number of records. Moreover, Italy borders the highest number of seas and includes the highest variability in habitat. As a consequence, the Italian records have been divided into zones: Ligurian Sea, Sardinia, Tyrrhenian Sea, Sicily, Isole Eolie, Isole Egadi, Pantelleria, Isole Pelagie, Ionian Sea, and Adriatic Sea.

The data bank contains a total of 596 or more records from the Mediterranean, of which 80 are doubtful. The distribution of the records is presented here in decreasing order. Italy tallied 251 or more, of which 23 or more are doubtful. These Italian records are divided as follows: Ligurian Sea, 27 or more, of which 3 or more are doubtful; Sardinia, 24; Tyrrhenian Sea, 70, of which 14 are doubtful; Sicily, 51, of which 2 are doubtful; Isole Eolie, 1; Isole Egadi, 19; Pantelleria, 4; Isole Pelagie, 9 or more; Ionian Sea, 5, of which 1 is doubtful; Adriatic Sea, 42 or more, of which 2 are doubtful; and unknown locations, 3. Croatia tallied 70 or more records, of which 1 is doubtful. There are 52 records from Spain, of which 5 are doubtful. Continental Spain tallied a total of 13 records, of which 3 are doubtful. Off the coast of Spain, the Isole Baleari tallied 38 records, of which 3 are doubtful. France tallied 52 or more records, of which 2 are doubtful. The French records were divided as follows: Continental France, 45 or more, of which 1 is doubtful; Corsica, 7, of which 1 is doubtful. Tunisia accounted for 42 or more of the records. Turkey tallied 24 records, of which 1 is doubtful. Malta tallied 14 of the records, of which 3 are doubtful, and Greece tallied 13 records. Other nations tallied very few records: Algeria tallied 5, Slovenia tallied 3, Montenegro tal-

4—Analysis of the Presence of Great White Sharks in the Mediterranean Sea

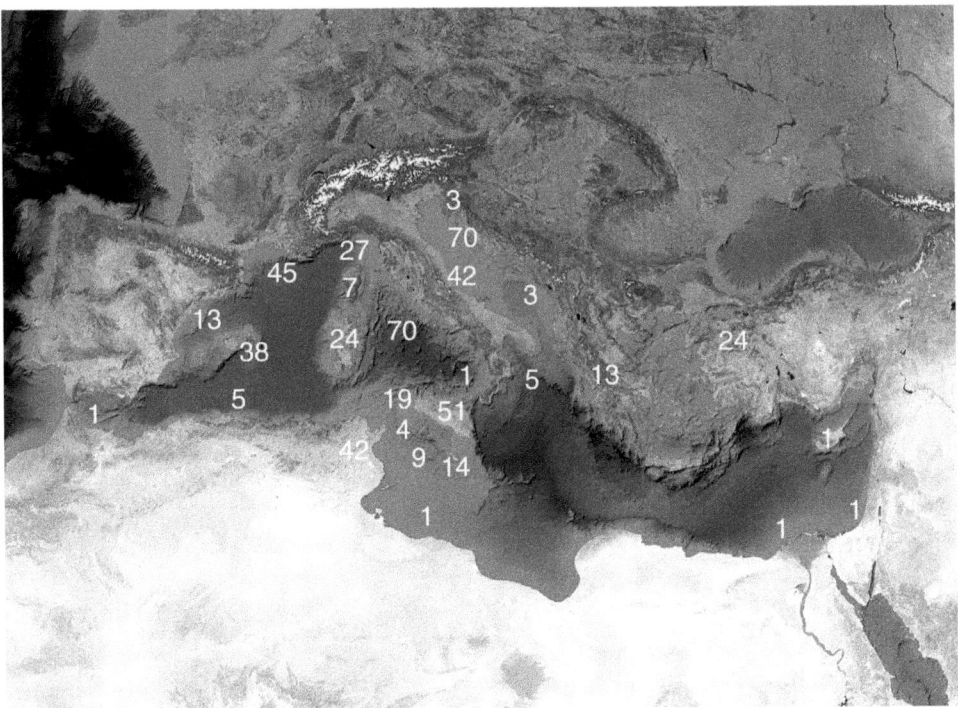

Map of white shark records with known location in the Mediterranean Sea (French Wikipedia).

lied 3, Cyprus tallied 1, Israel tallied 1, Egypt tallied 1, Libya tallied 1 and Morocco tallied 1. The nations bordering the Mediterranean that have no records include Monaco, Bosnia and Herzergovina, Albania, Syria and Lebanon. The data bank includes records without a location. Of those, 51 were recorded from the Mediterranean Sea, of which 45 are doubtful. The Adriatic Sea tallied 9 records without a location. In summary, combining the present and historical records, the number of records in decreasing order from the waters of the following nations are as follows: Italy, Croatia, Spain and France, Tunisia, Turkey, Malta and Greece. In the other nations, the number of records is minor. The records, both present and historical, can be presented based on geographical area. In decreasing order, the areas with the most records are the Adriatic Sea, Tyrrhenian Sea, the Channel of Sicily, Balearic Sea, Gulf of Lion, Ligurian Sea, Aegean Sea, Marmaric Sea and Ionian Sea. In the other waters, the number of records is minor.

It is important to recognize that the size and distribution of the Mediterranean white shark population has changed considerably over time. In some of the areas mentioned above, the white shark populations have decreased more than in others. In decreasing order, the areas where the size of the white populations have decreased most notably are the Adriatic Sea, especially in the Croatian waters (see page 203, "Fisheries"), the Gulf of Lion, Ligurian Sea, Balearic Sea and Marmaric Sea. Presently, the species appears to be rare everywhere in the Mediterranean. However, the areas where most records have been reported in recent years are the Tyrrhenian Sea, the Channel of Sicily and the Adriatic Sea.

The lack of records in some areas is not necessarily attributed to a small number of sharks but to different factors, especially a lack of communication between Italy (from where the study has been conducted) and some nations due to a language barrier or lack of cultural exchanges. In many areas there is a lack of shark researchers or ichthyologists available to document any sightings and captures of white sharks.

Identification

A photo-identification project of Mediterranean white sharks is now underway and may lead to some interesting results in the years to come. Similar photo-

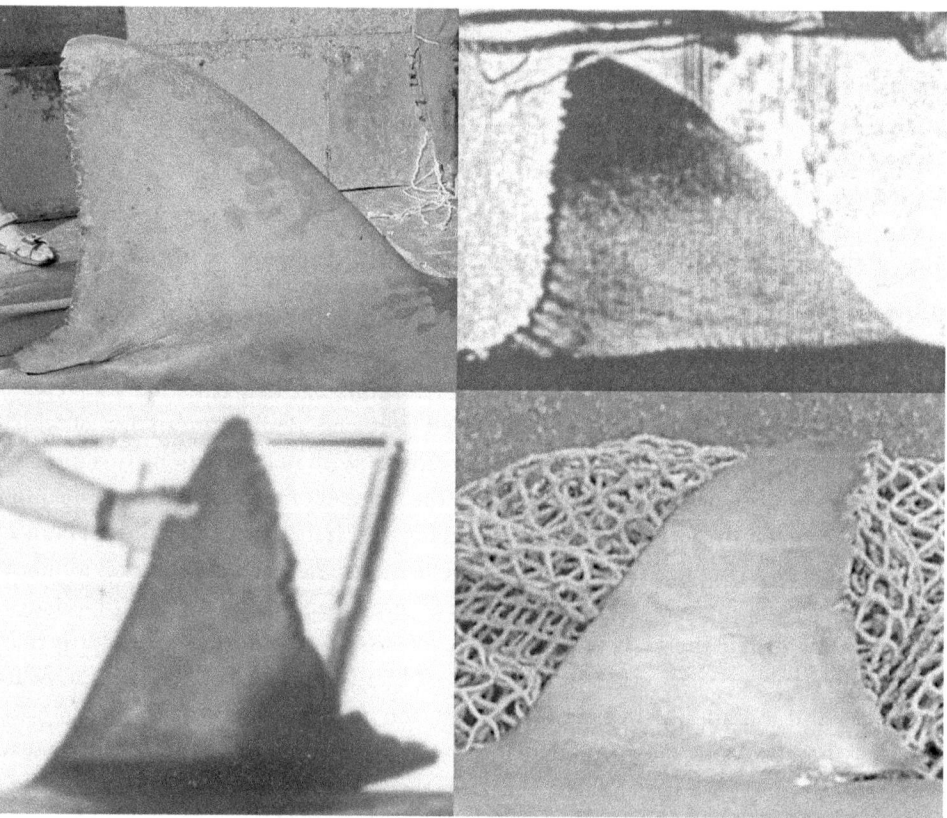

The only four specimens to which the finprinting technique was applied. From left to right: a female caught on October 18, 2009, off Mahdia, Tunisia; a female caught on April 17, 1987, near Filfla, Malta; a male caught on April 27, 2001, off Sidi Daoud, Tunisia, a female caught on April, 20, 2006, off Aras Dizra, Tunisia (Photographs by Mohamed Dahmouni, John Gullaumier, Walid Maamouri, Samuel P. Iglésias).

identification projects in other parts of the world have shown philopatry (site preference) in white sharks. For example, individual white sharks have been reported to revisit the waters off Gaudalupe Island in Mexico and the waters off Western Cape in South Africa on an annual basis (Domeier and Nasby-Lucas, 2007; Peschak and Scholl, 2006). Great white sharks can be reliably identified using each individual shark's unique trailing edge notch pattern, shape of the tip and pigmentation patches of the dorsal fin. This technique is named finprinting and is totally noninvasive because the sharks are identified by photographs or video frames (Peschak and Scholl, 2006). An attempt to apply finprinting in this study of the Mediterranean white sharks has been initiated. The major problem is that all of the numerous photos collected in the data bank were not shot by specialists but by occasional observers.

Unfortunately, finprinting could not be used on most of the recorded specimens. Most photos do not show the dorsal fin. Some photos showing the dorsal fin portray sharks killed well before the utilization of the finprinting technique that started in the 1990s. When the dorsal fin is shown in the photo, it is usually impossible to use it because it is out of focus, the resolution is low, it has been shot at an inadequate angle, or the dorsal fin is shown only in part and the part that can be seen is too small. In the end, there were only four sharks to which it was possible to apply the finprinting technique. These four Mediterranean specimens were then compared to over 1500 different individual white sharks identified in the largest existing white shark database, owned by shark researcher Michael C. Scholl of the White Shark Trust. Nevertheless, it turned out that these Mediterranean specimens were not among the specimens already identified in Scholl's database. The finprinting work is ongoing and with the improvements in digital photography and videography, it is likely that useful images will be available in the future.

Habitat, Movements and Seasonality

Great white sharks rarely swim very close to the shoreline. There are some exceptional cases in the data bank of white sharks swimming less than 50 m from the shore, which demonstrates that this does occur. The closest approach was a documented attack on a bather that occurred on June 1, 1963, only 3 m off Pithos or Micra Rock, Trikerion Island, Greece (Manolis Bardanis, pers. comm.). The next closest approach was a documented attack on a diver that occurred on October 6, 2008, only 10 m off the shore in the Bay of Mala Smokova, Lissa Island, Croatia (Branko Dragicevic, pers. comm.; Jakov Dulcic, pers. comm.). A near-shore attack occurred on a kayak on July 30, 1991, only 20 m off Portofino, Italy (Fergusson, 1996; Graffione, 1991). Another encounter only 20 m from the shore involved divers and occurred on June 20, 2002, off Cap Ferrat, France (De Maddalena and Révelart, 2008). On March 9, 1965, a shark was harpooned merely 40 m off Ganzirri, Sicily (Celona et al., 2001). Cases like these can occur in zones where the bottom drops very rapidly.

The shallowest depth of water recorded for a white shark in the Mediterranean was in waters that were estimated to be 80 cm deep, where an estimated 300 cm shark attacked a bather on June 1, 1963, off Pithos or Micra Rock, Trikerion Island, Greece (Manolis Bardanis, pers. comm.). The next shallowest depth of water recorded for a white shark in the Mediterranean was in waters that were estimated to be 1 m deep. This large shark was estimated at 560 cm in length. The sighting occured on March 9, 1965, where it was seen swimming off Ganzirri, Italy, chasing a school of mullet *Mugil* sp. The shark was reported touching the bottom with the ventral surface of its body, while its dorsal surface was out of the water (Celona et al., 2001).

This estimated 450 cm white shark was encountered on May 1988 off Numana, Italy (photograph by Fausto Fioretti).

Analysis of the data has shown that in the Mediterranean, as in other areas of the world, this predator frequents the area close to banks, islands, straits, and channels where prey is more abundant (Compagno, 2001; De Maddalena, 2002). Great whites are often found close to or at the surface, where they can be seen cruising slowly with the first dorsal fin and caudal fin upper lobe protruding out of the water. In the Mediterranean, great white sharks can be found in depths ranging from the surface to at least 130 m. This maximum depth was recorded off Marzamemi, Italy (Celona, 2002; Marino, 1965).

Great white sharks are nomadic and may spend relatively short periods of time at any given location. These sharks may exhibit philopatry, or site preference, which is the tendency to return to a specific location in order to breed or feed. The word philopatry is derived from the Greek, meaning "home-loving." Philopatry can manifest itself in several ways and can be applied to more than just the location where the animal was born. At least some individual sharks revisit these sites periodically, staying from a few days to several years (Compagno, 2001). Cases recorded in the Mediterranean of white sharks spending short periods of time at a given location or returning to it within a few days have been recorded for the following locations and periods: in August 2001 off Torre delle Stelle alias Capo Torre Finocchio in Sardinia, Italy; in June 1721 off Ponte della Maddalena in Napoli, Italy; in June 1978 in the Golfo di Venezia; in August 1938 off Enfola in Isola d'Elba; in September 1956 on the Secca del Faro off Circeo; in

August-September 1962 and then in the early summer of 1963 and again in November 1964 on the Secca del Quadro off Circeo, Italy; in January-May 1989 in the Canale di Piombino; in May 1990 in the Stretto di Messina; in the end September 1986 between Rimini and Pesaro; between the end of August and early September of 1934 off Rijeka, Kraljevica and Susak, Croatia; in 1993 off Paphos, Cyprus. In some cases, it is thought that the same shark revisited the same site after a long period of absence. This kind of case has been reported from the Mediterranean, in the Golfo di Baratti, on the Secca del Quadro off Circeo, and between Rimini and Pesaro, Italy. However, these cases can not be confirmed, since in none of them has there been any definitive feature to identify an individual shark.

In the Mediterranean, there are pronounced periodic increases in white shark abundance in some regions, which are correlated with Atlantic bluefin tuna *Thunnus thynnus* concentrations. Atlantic bluefin tuna are divided into two stocks. The eastern stock has a reproduction area that includes the Mediterranean, and a western stock with a reproduction area in the Gulf of Mexico. In the Mediterranean, the Atlantic bluefin tuna reach sexual maturity at an age between three and four years old, with a corresponding average length of 105 cm. The large mature eastern Atlantic tuna that approach European coasts in May pass through the Gibraltar Strait and enter the Mediterranean, navigating relative to the Atlantic current to reach the preferred reproduction areas, located in the western, central and eastern parts of the Mediterranean basin. The bluefin tuna reproduce when the water temperature reaches or exceeds 24°C, and in the Mediterranean this generally occurs at different times depending on the location. The conditions are right for spawning around mid May in the Levantine Sea, around mid June in the central Mediterranean (Tyrrhenian Sea and Maltese waters) and around June in the Balearic Sea.

After spawning, the tuna can leave the Mediterranean, moving back to the Atlantic through the Gibraltar Strait. The sexually mature bluefin will return to the Mediterranean the following year to spawn again. Other bluefin tuna are residents, remaining in the Mediterranean near the birth area (homing). These resident tuna frequent the waters of Isole Eolie, in the Stretto di Messina, in the Golfo del Leone, around the Bocche di Bonifacio and in other areas where there are high concentrations of food fish, which correspond to areas of high primary productivity (Gregorio De Metrio, pers. comm.; Dino Levi, pers. comm.). It is hypothesized that there may be two different Mediterranean tuna populations based on possible genetic differences between tuna from the Tyrrhenian Sea and those from the Levantine Sea. The eastern Mediterranean population may be separate from the western and central Mediterranean population (Carlsson et al., 2004).

It is not surprising that in the Mediterranean, white sharks are more frequently recorded from May to September and that the areas where these large predators are more abundant coincide with those areas where Atlantic bluefin tuna are more abundant. This is particularly evident from the number of white sharks unintentionally captured in tuna traps (see page 203, "Fisheries"). White shark movements in the Mediterranean are therefore tightly linked to those of Atlantic bluefin tuna. From the data collected, there seems to be no set migration pattern of the white sharks in the Mediterranean. They apparently adapt to the variety of routes taken by the tuna. Rather than discussing migration routes, it is probably better to refer to various sites of aggregation, where the sharks congregate at certain times of the year. However, from the data collected in this study, there is no reason to believe that great white sharks follow tuna in their migratory route through the Gibraltar Strait. It is certainly possible that this may happen occasionally, but the lack of records from the Alboran Sea seems to suggest that the white sharks tend to remain in the Mediterranean.

We can hypothesize the existence of a Mediterranean white shark population, partially separate from the Atlantic population. Further studies based on the tagging of Mediterranean white sharks and genetic studies of the species in these waters are needed to validate the hypothesis. However, the results of this study seem to confirm recent findings suggesting that Mediterranean white sharks probably constitute a sink population established at a time of extreme climate change. Gubili et al. (2010) sequenced the mitochondrial control region of four Mediterranean white sharks and found that they were very different to the Atlantic specimens and more like white sharks from Australia and New Zealand. This genetic study suggested that great white sharks in the Mediterranean may have first arrived from the seas around Australia about 450,000 years ago, a time of interglacials. The climatic instability of the Pleistocene may have induced navigational errors, with sharks following an Agulhas ring or eddy remarkably stronger than the contemporary phenomenon, directing animals north along the African coast. This could have led a few pregnant females to an entrapment in the Mediterranean, where then they gave birth to their pups and, as a species exhibiting strong natal female philopatry, the new population would have remained in the Mediterranean because they return to spawn where they were born.

Diet

There are 111 Mediterranean records that include information on the diet of the great white sharks. These reports may include stomach contents from dead specimens, sharks observed preying on live animals, sharks observed scavenging carcasses, evidence of attempted or successful predation observed on wounded animals (e.g., a bite on a live or moribund animal), and evidence of scavenging on carcasses (e.g., chunks of flesh removed from a carcass). Of these categories, observed predatory events are very rare, as, excluding attacks on humans, they were observed in only three cases.

Data about diet from Mediterranean records cannot be considered exhaustive, because white shark stomach contents are reported mainly when the contents are unusual

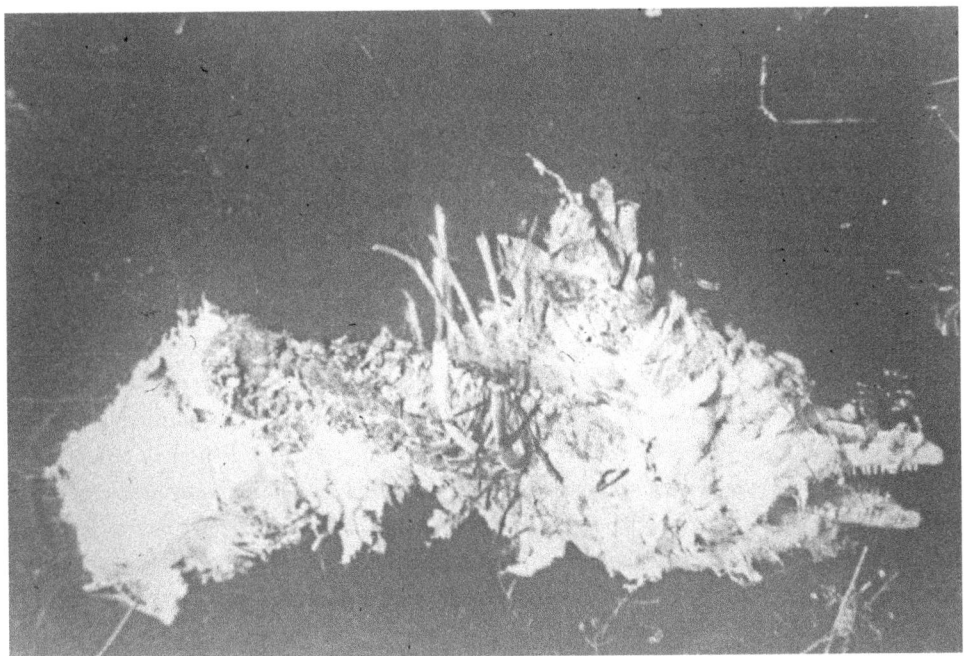

Remains of a bottlenose dolphin *Tursiops truncatus* that were regurgitated by an estimated 500 cm white shark encountered on June 7, 1978, in the Golfo di Venezia, Italy (courtesy Luigi Cavaleri).

or impressive, such as when they include large prey, especially if the prey was consumed whole. Unusual items such as human clothing get reported, whereas sharks with normal prey are less likely to be reported. Thus, the reported stomach contents may not represent the typical diet of Mediterranean white sharks (De Maddalena, 2002).

In the Mediterranean, fish form the main diet items of the great white shark. Feeding on fish was recorded in 48 cases. Of these, 6 refer to unidentified fish, 36 refer to bony fish, and 5 refer to cartilaginous fish. The 37 records of feeding on bony fish are divided as follows: 19 refer to unidentified tuna *Thunnus* sp. (including 1 doubtful case), 2 refers to Atlantic bluefin tuna *Thunnus thynnus*, 1 refers to albacore *Thunnus alalunga*, 3 refer to Atlantic bonito *Sarda sarda*, 1 refers to Atlantic mackerel *Scomber scombrus* (doubtful), 6 refer to swordfish *Xiphias gladius*, 1 refers to mullet *Mugil* sp., 1 refers to European pilchard *Sardina pilchardus*, 1 refers to grouper (doubtful), 1 refers to scorpionfish *Scorpaena* sp. and 1 refers to common dentex *Dentex dentex*. The 5 records of feeding on cartilaginous fish are divided as follows: 1 refers to a ray (manta or ray), 1 refers to a shortfin mako shark *Isurus oxyrinchus*, 2 refer to thresher sharks *Alopias* sp. and 1 refers to a blue shark *Prionace glauca*. In some cases, the prey fish were not taken naturally by the shark. Some of the swordfish and tuna were stolen from fishermen while they were hooked on recreational or commercial fishing equipment or tied to the side of the boat (see page 194, "Scavenging").

Marine mammals are an important food source for white sharks. The data bank has 42 cases (3 are doubtful) of feeding on marine mammals. Of these, 1 case refers to a pinniped and 41 cases refer to cetaceans. The Mediterranean monk seal *Monachus monachus* is the only pinniped present in the area. There was just one record of a white shark feeding on a monk seal, dated from 1956 (Damonte, 1993). Due to the present day scarcity of this pinniped, it is a possible prey but probably not a significant source of prey for Mediterranean white sharks (De Maddalena and Révelart, 2008).

Cetaceans represent an important food item for great white sharks (Arnold, 1972; Compagno, 1984; Long and Jones, 1996). The 41 records of feeding on cetaceans in the data bank are divided as follows: 22 refer to unidentified dolphins, 1 refers to a common dolphin *Delphinus delphis*, 3 refer to striped dolphins *Stenella coeruleoalba*, 6 refer to bottlenose dolphins *Tursiops truncatus* (including 2 cases that are doubtful), 1 refers to a Risso's dolphin *Grampus griseus*, 2 refer to harbor porpoises *Phocoena phocoena*, 2 refer to a sperm whale *Physeter macrocephalus*, 2 refer to a fin whale *Balaenoptera physalus* (which are a type of baleen whale and one case is doubtful) and 2 refer to unidentified baleen whales. In total, there were 35 cases of feeding on dolphins (including 2 that are doubtful) recorded in the data bank. Based on the data, dolphins must be considered an important and frequent food item in the diet of the Mediterranean white sharks. Although white shark predation on dolphins was never observed directly, it was shown that this large predator does not just feed on dolphin carcasses. An estimated 350 cm bottlenose dolphin was photographed near Lampedusa, Isole Pelagie, Italy, showing two fresh white shark bites in the dorsal region, which proves that the great white shark does prey on live dolphins in the Mediterranean (Celona et al., 2006). Some of the bite marks on dead dolphins are consistent with a swimming animal that was attacked, like bites forward of the tail. The harbor porpoise was once considered by some authors as common in some Mediterranean areas. There are two cases where harbor porpoises were found in white shark stomachs, both cases dated at the end of the 19th century. Currently, the small harbor porpoise is extremely rare, making it a possible prey for Mediterranean great whites but probably just an occasional meal (De Maddalena and Zuffa, 2008).

Six cases of scavenging on large cetaceans were recorded in the data bank, including 4 dead or dying baleen whales (including 1 case that is doubtful) and 2 sperm whales. In 2 of the cases of scavenging on baleen whales, the source reported the whales to be blue whales *Balaenoptera musculus,* but it is likely that the identification was in error since the blue whale is not a Mediterranean species and has never been reported from this area. Since the source reported that the baleen whales measured 18 m and 21 m, respectively, they were most likely fin whales *Balaenoptera physalus*, which is a common species in the Mediterranean Sea and can be mistaken for a blue whale.

Concerning marine reptiles, there are 11 cases of feeding on sea turtles recorded in the data bank. In 4 of these cases, the species of sea turtle was not reported, while in 6 cases it was a loggerhead sea turtle *Caretta caretta* and in 1 case it was a green sea turtle *Chelonia mydas*. In 8 of the cases, the turtles were found in the stomach of the captured white shark. In 2 cases, the sea turtle carcasses were found bearing white shark bites and in a single case, a live sea turtle was found bearing signs of a predatory attack by a white shark. There are two cases in the data bank reporting feeding on molluscs. One case refers to unspecified molluscs and the other refers to squid. There are 5 cases of feeding on a bird. One refers to some feathers found in the stomach of a white shark, another to a seagull found in the stomach of a shark, and another to albatrosses found in the stomach of a shark. The other two cases refer to predation on seagulls, but one is doubtful.

Besides the more typical prey items, there were 10 cases recorded of feeding on terrestrial mammals. The terrestrial animals included dogs *Canis lupus familiaris*, cats *Felis silvestris catus*, goats *Capra hircus*, lambs *Ovis aries*, pigs *Sus domesticus*, calves *Bos taurus* and horses *Equus caballus*. In most of these cases, it is assumed that the white sharks fed on carcasses of these mammals (the items found include even a ham bone). However, it is possible that in some cases, a terrestrial mammal was preyed upon as it was swimming at the surface, which has been reported in other seas (Lineaweaver and Backus, 1979).

Comparing the diet of the great white sharks from the Mediterranean with those inhabiting other areas, there are notable differences. For example, the examination of the stomachs of 591 white sharks caught along the coast of Natal, South Africa, showed that 41.1 percent of the contents consisted of elasmobranchs, 34.7 percent of bony fish and 29.0 percent of marine mammals (Cliff et al., 1989). It is likely that the size of the elasmobranch populations in the Mediterranean Sea were much smaller than those of the South African waters. It is also certain that many Mediterranean elasmobranch populations were almost wiped out by the commercial fisheries in the last 40 years (De Maddalena and Baensch, 2005).

It is well known that the diet of the great white shark varies from area to area (De Maddalena, 2002). Many sharks are opportunistic feeders, meaning that they are versatile and able to utilize diverse food sources depending on the availability of each food type. When their favorite prey is scarce, they will eat other species that are locally more abundant. A particularly common species or one that is easily captured may dominate the diet of opportunistic sharks (De Maddalena, 2008). In the Mediterranean, there is not a high abundance of elasmobranchs, nor is there a sufficient number of pinnipeds to support a population of white sharks. So, the Mediterranean great white sharks base their diet primarily on bony fish, especially tuna, and secondarily on cetaceans, especially dolphins. Sea turtles are also a prey item for larger sharks. To a lesser extent, Mediterranean white sharks feed also on elasmobranchs, especially on other sharks, and on molluscs and sea birds.

Previous studies have shown that the diet of the great white shark is closely related to its size and age. Young white sharks have teeth that are more narrow, finely serrated or almost smooth. These teeth are better adapted for grasping than for cutting, hence the young individuals feed on smaller prey which they usually swallow whole. Larger sharks have teeth that are broader and heavily serrated, adapted for cutting prey into smaller pieces or for neatly excising a mouthful of flesh from large prey. Large prey become increasingly important in the diet of sharks as they grow larger. Analysis of prey type in relation to great white shark size shows that small sharks measuring less than 300 cm in length feed primarily on fish, while larger sharks have been identified as important predators and scavengers on marine mammals (Compagno, 2001). This aspect of the white shark's diet was partially confirmed by this study, because all diet-related data collected from the Mediterranean sharks was from specimens measuring over 400 cm. Dolphins and tuna were found to be a food item of white sharks measuring 400 cm or over in size. Sea turtles were found in specimens measuring 500 cm or over, but it is not known if these turtles and dolphins are even eaten by white sharks measuring less than 400 cm.

The great white shark is one of the more omnivorous sharks, and it can eat almost anything. Great white sharks are very curious animals. They investigate and sometimes strike, bite and swallow foreign objects, including unusual food and inedible items, and indeed great white sharks are known to consume inedible items (De Maddalena, 2002; Lineaweaver and Backus, 1979). Collier et al. (1996) demonstrated that great white sharks attack inanimate objects of a variety of shapes, sizes and colors, none resembling the shape, size or color of their typical prey. Consequently, it has been suggested that these predators are determining the suitability of the prey as food in these cases.

These animals have a strong tendency to strike non-prey objects and may be using them for target practice. While white sharks are attracted to metal items due to their electroreception capability, they swallow plastic items that give off no electric field. According to some researchers, heavy objects could be ingested to solve buoyancy control problems. The sometimes indiscriminate feeding habits of great white sharks have been well documented in the Mediterranean. There are 18 cases of sharks that swallowed inedible items, which are recorded in the data bank. These items include stones, a metallic cable, a boat chain, 31 hooks, pants, boots, pieces of canvas, plastic bottles, plastic bags, shoes, socks, a wool bodice, baskets, a raincoat, 2 or 3 coats, an automobile license plate, a small board of cork, a plastic container, a box containing garbage, clothes, a wig, a broomstick, a washing machine drum, a doll, one or more 1 kg meat cans still sealed, a 5 kg can, and other unidentified garbage. There is even a case of a shark swallowing a man cloaked in a suit of armor during the Middle Ages.

Predatory Tactics

Great white sharks play an important ecological role in marine communities. As predators, they are fundamental instruments of natural selection. In general, great white sharks are solitary hunters (Compagno, 2001; De Maddalena, 2008). In the Mediterranean, most of the records from the data bank are of individual sharks. There are 18 records (including 2 doubtful cases) of white sharks that were encountered in pairs or in groups of 3 individuals. As in other waters, it is clear that in the Mediterranean, great white sharks are solitary hunters.

The great white shark is a stalking hunter, thus the success of this predator depends on both speed and the element of surprise. It is thought that the great white shark would have difficulty capturing a healthy, fast-swimming prey if the prey was aware of the predator's presence (Strong, 1996). Great white sharks attack the fast-swimming animals on which they feed suddenly and violently. The victim typically does not see the shark until it is too late. It is overwhelmed by the unexpected assault and the violent force with which the attack is executed. This predatory tactic enables the shark to obtain its meal with minimal risk of injury as well as minimal energy expenditure. These sharks have been reported to attack seals, sea lions, sea otters, dolphin, tuna and humans using this method (Ames et al., 1996; Collier, 2003; Levine, 1996; Long et al., 1996; Long and Jones, 1996; Miller and Collier, 1980; Tricas and McCosker, 1984; West, 1996). This attack strategy where the unsuspecting victim did not see the shark before the attack was also recorded in the Mediterranean.

Great white sharks can be oriented horizontally or vertically when they approach their potential prey. The predator uses its heavy mass and speed to violently ram and stun the prey. During vertical approaches, the great white shark attacks from below, by swimming from depths as deep as 17 m and moving on an upward line that is 45°–90° oriented from the prey (Strong, 1996). The kinetic energy of the shark is proportional to the mass of the shark and to the square of the velocity, which allows a large, fast-moving shark to inflict heavy damage on large prey. Prey have been observed being propelled out of the water by the force of the shark's impact, rendering the prey disoriented and therefore incapable of resistance.

There are 8 cases of a white shark feeding on a dolphin recorded in the data bank in which the body region of the bite was reported. In 6 cases, the bite region was the urogenital region, in 1 case it was the dorsal region, and in 1 case the caudal fin was severed (Celona et al., 2006; Centro Studi Cetacei, 1988, 1989; Mauro Cottiglia, pers. comm.; Morey et al., 2003; Marco Zuffa, pers. comm.). These reports are consistent with what was previously observed in other seas (Long and Jones, 1996). In order to avoid detection by dolphins, great white sharks approach these mammals from below, above or behind, because odontocetes have an anteriorly directed sonar and a lateral visual field. The caudal peduncle, urogenital region, abdominal area, and dorsum are bit more often, while wounds to the head and flanks are less common (Long and Jones, 1996). Shark bite scars are more often observed on the dorsum of live cetaceans since a bite in this area is less likely to incapacitate the dolphin, providing it a higher probability of surviving the attack (Celona et al., 2006; Long and Jones, 1996). This notion is supported by the case recorded in the Mediterranean, on May 5, 2006, of an estimated 350 cm bottlenose dolphin *Tursiops truncatus* encountered near Lampedusa,

This estimated 350 cm bottlenose dolphin *Tursiops truncatus*, showing two fresh white shark bites in the dorsal region, was photographed on May 5, 2006, near Lampedusa, Isole Pelagie, Italy (photograph by Emiliano D'Andrea/Necton Marine Research Society).

Isole Pelagie, showing two fresh white shark bites on the dorsal region (Celona et al., 2006).

Scavenging

Not only is the great white shark a predator of live animals, but it is also a scavenger. An important distinction must be made between preying on a live animal and eating a dead animal. Scavengers feed on dead organisms. Feeding on dead animals is very advantageous because it often provides enough energy to sustain a shark for long periods with minimal energy expenditure (De Maddalena, 2008). Large marine animal carcasses are known to attract great white sharks. As stated previously, great white sharks are solitary hunters but may gather in relatively high numbers of up to ten individuals or more at a large cetacean carcass (Compagno, 2001; De Maddalena, 2008). In the Mediterranean, great white sharks scavenge on a wide variety of dead animals. Fish processing plants and slaughterhouses that dump their wastes into the sea sometimes had white sharks swimming very close to the plant. This is documented in the case of a 550 cm white shark in 1953 that ventured close inshore in Favignana, Isole Egadi, Italy, swimming into the bay where the old tuna processing plant was located,

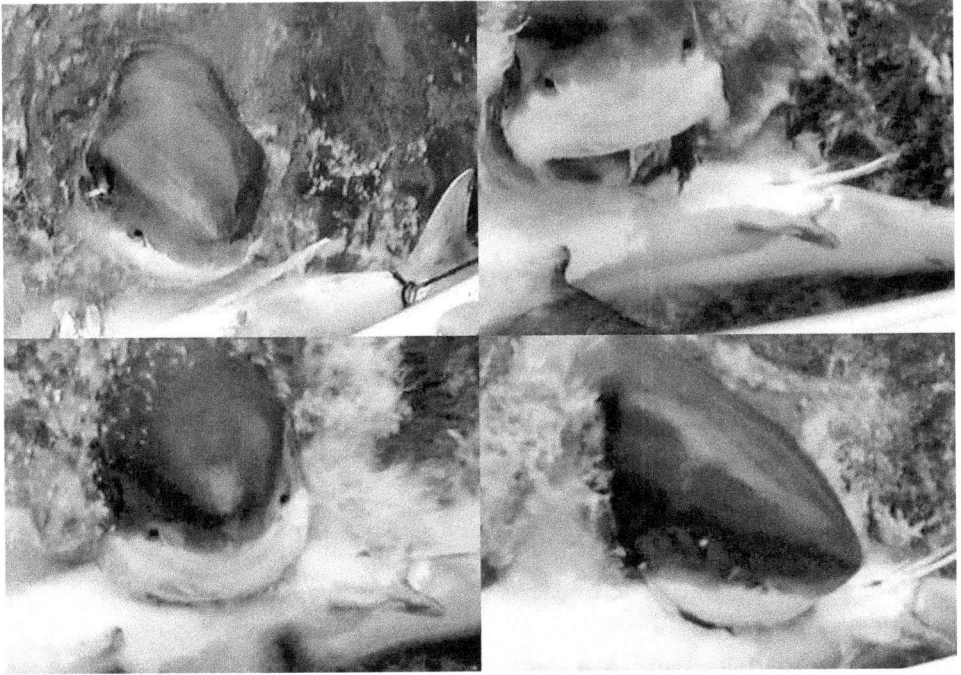

An estimated 500–600 cm white shark scavenging a common thresher shark Alopias vulpinus on August 27, 1998, off Senigallia, Italy (photograph by Stefano Catalani).

likely attracted by the great amount of blood in the water (Gioacchino Cataldo, pers. comm.; Giudici and Fino, 1989; Mojetta et al., 1997; Salvatore Spataro, pers. comm.). Another case occurred some years before 1878, when an estimated 500 cm shark ventured close inshore in Syros, Cyclades Islands, Greece, to eat some leather that was placed in the water near the shore close to a tannery (Heldreich, 1878). Dead whales provide a source of high-energy blubber, which is a prime food source for white sharks. There were six records in the data bank showing evidence of scavenging on whale carcasses. Scavenging is often the reason terrestrial animals are found inside sharks (see "Diet," page 188).

Fishery catches are important food sources for great white sharks. Fish are sometimes torn from hooks by the sharks. These predators are often cautious when investigating hooked baits. Some great white sharks that feed on hooked or netted fish are captured while feeding. Even if the shark is not hooked, it can be trouble for the fishermen. As testified to by some of the cases recorded from the Mediterranean, when a great white shark approaches a fishing boat and starts to eat a hooked fish, someone pulling the fish away from the shark and aboard the boat can provoke an aggressive reaction, where the shark may bite the hull or in rare cases seriously damage the boat. A hooked white shark or entangled white shark is even more trouble for the fishermen. When white sharks swim into a gill net, they sometimes become entangled in the mesh and by their efforts to escape create costly damage for the fishermen. Often, the great white sharks that were caught in the nets were pursuing the same fish as the fishermen and became entangled together with the potential prey. Therefore, scavenging by great white sharks may be a reason for simultaneous capture. Moreover, the stomachs of these sharks often contain tuna and swordfish probably torn from hooks. There are some cases in the data bank where a white shark stole a swordfish tied to the side of a fishing boat and in one case stole a captured thresher shark that was hanging off the side of the boat.

Great white sharks will take advantage of catastrophes, whether caused by humans or by nature, to feed on wounded, dying or dead animals. The scent of blood, loud sounds, and movements made by dying individuals are detected by sharks from a long distance. Catastrophes such as seaquakes, disease outbreaks, underwater explosions, shipwrecks and airplane crashes soon attract numerous sharks. In Sicily, Italy, a terrible earthquake took place on December 20, 1908. There was a huge seaquake that caused enormous destruction in Messina. The large tsunami caused by the seaquake caused numerous deaths and disappearances. During this period, a great white shark was found stranded in Messina and its stomach contained the leg of a woman. One month later, on January 26, 1909, another great white shark was caught off Augusta, approximately 100 km south of Messina. This shark had the remains of at least three people (a man, a woman and a child) in its stomach. In both instances, the victims undoubtedly were people drowned by the tsunami.

Predators and Parasites

Great white sharks are at the top of the biomass pyramid and have few enemies. Only a few creatures prey on great white sharks. These include humans and other great white sharks, but probably only under certain circumstances, such as when a white shark is hooked or injured. Killer whales *Orcinus orca* (Linnaeus, 1758) also prey on great white sharks but only in very rare cases (Pyle et al., 1999). These cetaceans are virtually absent from the Mediterranean. Nevertheless, great white sharks are hosts to numerous parasites such as copepods, trematods, and cestodes. These parasitic organisms live in or on the great white shark in order to obtain sustenance from them. Multiple infections by parasites are common and can cause serious lesions, sometimes resulting in severe diseases (De Maddalena, 2008). Parasites of the great white shark observed in the Mediterranean include the copepods *Echthrogaleus coleoptratus* (Guerin-Meneville, 1837) and *Nemesis lamna* (Risso, 1826), the cestode *Tetrarhynchus megacephalus* (Rudolphi, 1819), and the trematod *Distomum continuum* (Brian, 1906; Tortonese, 1956). The copepods typically attach somewhere on the exterior of the shark—like the trailing edge of the fins, in the gum area of the mouth or in the gill slits—while the cestodes and trematods are internal parasites.

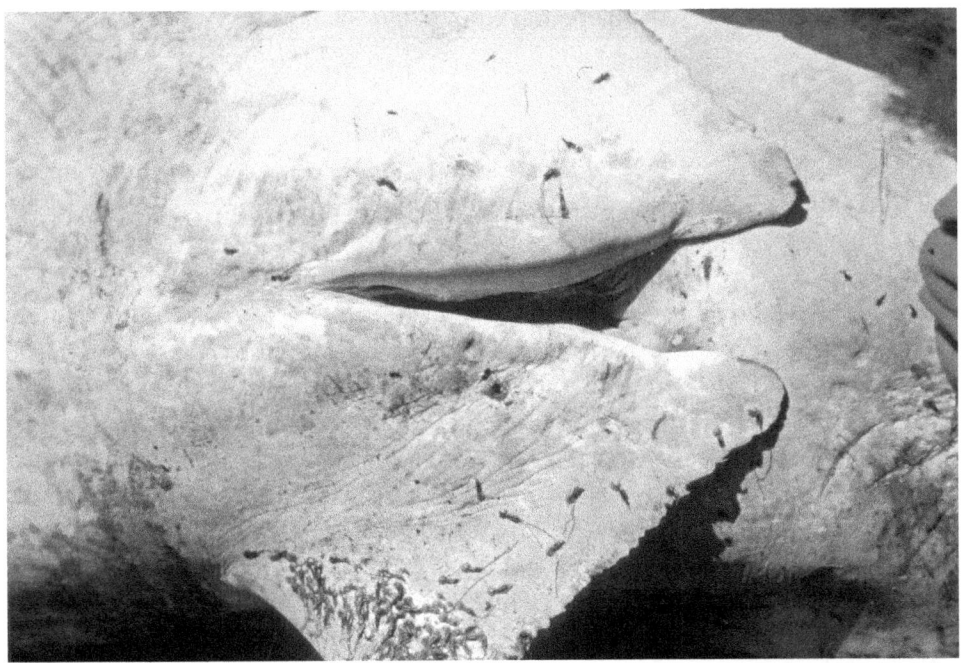

Top: Copepod parasites in the cloaca region of a 535 cm female white shark caught on May 8, 1987, off Favignana, Isole Egadi, Italy (courtesy Gioacchino Cataldo). *Bottom:* This estimated 700–800 cm white shark encountered on September 9, 2002, off Porto San Giorgio, Italy, was accompanied by a group of pilot fish *Naucrates ductor* (photograph by Glauco Micheli).

Mutualism

Like all large sharks, great white sharks are sometimes accompanied by remoras and pilot fish. Remoras are bony fish of the family Echeneidae. In the Mediterranean, the prominent species is the common remora *Remora remora* (Linnaeus, 1758). Remoras have a dorsal suction disk formed from their modified dorsal fin, which they use to attach themselves to sharks, mantas, marine turtles and other large pelagic creatures. They use this suction disk only when the large animal changes direction or slows down. The pilot fish *Naucrates ductor* (Linnaeus, 1758) is a bony fish of the family Carangidae that is often associated with cartilaginous fish, other large bony fish and marine turtles. The relationships between pilot fish and great white sharks and between remoras and great white sharks are cases of mutualism, because each organism benefits from the other. Pilot fish and remoras benefit from the relationship with a great white shark by eating the scraps of food or excrement and parasites, as well as by riding the shark's bow wave. Conversely, great white sharks benefit from the relationship with remoras and pilot fish by being cleaned of parasites. Observations seem to indicate that pilot fish are much more common companions for great white sharks than the remoras, which are observed less often on the body or in the proximity of these large predators (De Maddalena, 2009).

Remarkably, remoras were never observed on the body of great white sharks or in proximity to these predators in the Mediterranean. Pilot fish seem to be just sporadic companions of Mediterranean white sharks, considering that their presence close to white sharks was observed in only 5 cases. All white sharks that were seen to be accompanied by pilot fish were at least 320 cm-long. The maximum number of pilot fish seen near a single white shark was 20–30 individuals, which was recorded on September 9, 2002, 50 miles from Porto San Giorgio. A pilot fish that was observed swimming near a white shark recorded on April 29, 2000, 1.5 miles off Faro di Piave Vecchia, Italy, was collected and is preserved at the Museo Civico di Storia Naturale in Venezia, Italy (Luca Mizzan, pers. comm.).

Attacks on Humans

As a result of several data collection programs, a substantial amount of information about historical and recent shark attacks has been collected. These programs focus on research specific to shark attacks on humans. Two such programs are the Global Shark Attack File (GSAF) and the International Shark Attack File (ISAF, a compilation of shark attacks worldwide based at the Florida Museum of Natural History in Gainesville, Florida, USA). The Italian Great White Shark Data Bank includes a compilation of great white shark incidents that occurred in the Mediterranean.

Author De Maddalena collaborates with the GSAF as a regional investigator for the Mediterranean Sea. The GSAF gathers data on shark attacks worldwide and is maintained by the Shark Research Institute (SRI) based in Princeton, New Jersey, USA. The GSAF quantitative data is available to the medical profession, the scientific community, the media, and the general public in order to provide accurate and current data on shark and human interactions. In addition, the file gathers medical data in a format that can be utilized by physicians and surgeons called upon to treat victims of shark attacks.

According to ISAF, worldwide there are 70 to 100 shark attacks annually, and only 5 to 15 are fatal. It is quite surprising that so few incidents occur, considering the wide distribution of the most dangerous sharks and the large number of people who swim and dive in the seas of the world. However, the total number of shark attacks around the world per year is actually unknown, and the recorded incidents represent only a portion of the actual total. In fact, in many areas of the world, shark attacks go unrecorded. The lack of information on most white shark attacks that occurred in the Mediterranean is quite surprising. Often, the data collected is extremely vague and questionable. Almost all attacks recorded in the study area occurred a very long time ago, making it very hard or impossible to find reliable sources and especially eyewitnesses.

Sometimes great white sharks scavenge on human cadavers in the Mediterranean. Scavenging on corpses is not considered an attack. Therefore, in the analysis of attacks on humans from the Mediterranean, the 13 cases of human cadavers found in the stomachs of great white sharks were not tallied as attacks. It is usually impossible to find indications that the stomach contents were taken through active predation rather than postmortem scavenging. Nevertheless, for 3 of these 13 cases it is certain that they were the result of scavenging. In one case, an entire human corpse sewn into a sailcloth with cast-iron weights was clearly a seaman buried at sea (Biagi, 1989, 1995). In two other cases, the victims were undoubtedly drowned by a huge seaquake that caused enormous destruction and resulted in 120,000 deaths and disappearances in Messina, Italy (Berdar and Riccobono, 1986; Condorelli and Perrando, 1909; Munthe, 1928).

In total, 55 white shark attacks were recorded in the Mediterranean, including 13 that must be considered doubtful because there is some doubt over the exact identity of the species or because the case itself is considered doubtful. In 4 cases, the shark identity was more certain because the shark left some teeth or tooth fragments embedded in the boat and in 1 case in the body of the victim. In 3 of these cases, the teeth were examined by marine biologists who confirmed the identity of the shark responsible for the attack as a white shark (Mauro Cottiglia, pers. comm.; Branko Dragicevic, pers. comm.; Jakov Dulcic, pers. comm.).

Of the 55 cases, the attack was fatal in 19 cases, and nonfatal in 19 cases (including 3 doubtful cases). For the other 17 cases (including 10 doubtful cases), the outcome of the attack was not reported. The percentage of fatal attacks recorded in the Mediterranean is 34.5 of the total, which is higher than the worldwide percentage of 26.1

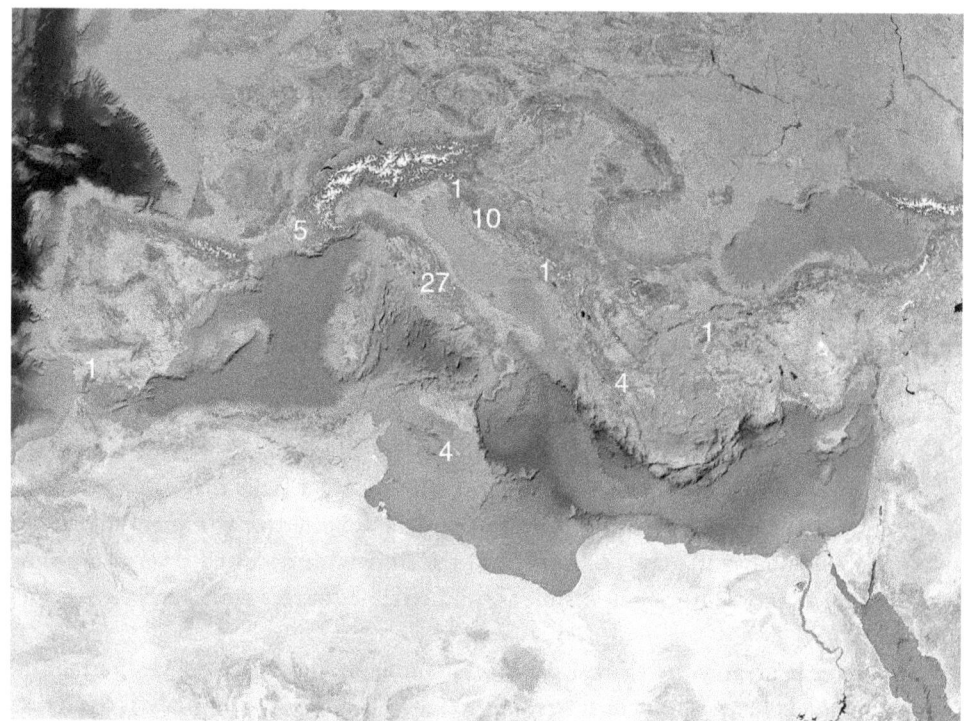

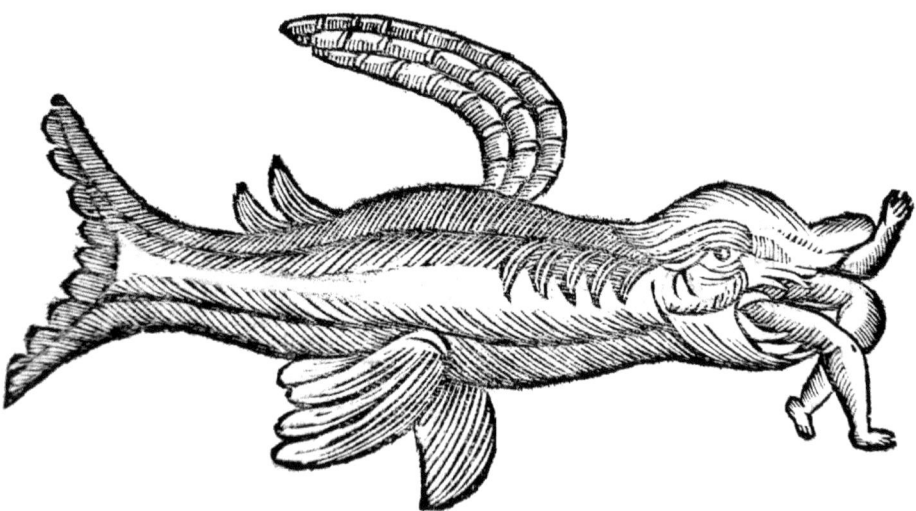

Top: Map of white shark attacks in the Mediterranean Sea (French Wikipedia). *Bottom:* Illustration portraying a 527 cm female white shark caught on June 6, 1721, off Napoli, Italy. A few days before the capture, a person was attacked and eaten by this shark off the beach called Ponte della Maddalena, in Napoli (illustration from Ricciardi, 1721).

(Burgess and Callahan, 1996). Because the outcome for 18 attacks is unknown, the percentage is likely higher. One hypothesis to explain this difference is that the Mediterranean white sharks responsible for the attacks were very large and the damages inflicted by the attacks had to be devastating.

In general, an attack on humans ends after the initial contact, and the great white shark does not eat or kill the victim. This was found to be true for most recorded attacks in the Mediterranean. Most attacks result in significant blood loss but not massive consumption by the shark. It is true that the human–great white shark attack fatality rate is relatively high compared to that for other species, but death is usually a result of blood loss and shock. Human beings are not a usual part of any shark diet, great white shark included. It is presumed that the great white sharks do not regard humans as food and that the large majority of attacks are not motivated by hunger. All white shark attacks on humans can be divided into two categories—provoked and unprovoked. When the behavior of the victim elicits aggression in the shark, the attack is considered to be provoked. The data collected from the Mediterranean, just as elsewhere in the world, demonstrates that white sharks are typically not naturally aggressive toward people. In the Mediterranean, only 2 cases (including 1 doubtful) were reported as provoked attacks, while 19 cases (including 1 doubtful) were reported as unprovoked attacks. Unfortunately, for the other 34 attacks recorded (including 11 doubtful), it is not known whether the attack was provoked or unprovoked. Examples of provoking aggression in the shark are hitting the shark with naked hands, an oar or a spear. There are cases in the data bank where a white shark was encountered by divers and was not aggressive.

Unprovoked attacks on humans by great white sharks may be attributed to various causes, including a form of play or hunting practice, perception that the human is a threat, defense, interference with feeding, determination of the suitability of the victim as potential food, and mistaking the human as usual prey (Collier, 2003; Cousteau and Richards, 1992; De Maddalena, 2008; Martin, 2003). Tricas and McCosker (1984) have suggested that great white sharks mistake surf board silhouettes for pinnipeds. Burgess and Callahan (1996) reported that a high percentage of victims wear black wet suits or clothing similar to the dark coloration of many marine mammals, thus they look like the usual prey. In a feeding study, Strong (1996) observed that these sharks prefer a target shaped like a seal when presented simultaneously with a square target. However, Collier et al. (1996) demonstrated that great white sharks also attack inanimate objects of a variety of shapes, sizes and colors, none resembling the shape, size or color of a pinniped. Consequently, they have suggested that these predators may be determining the suitability of the prey as food. As it was pointed out by De Maddalena (2009), great white shark attacks on humans occur even in geographical areas where pinnipeds are not a significant prey item for the great white shark, which is true for the Mediterranean Sea. Presently, the scarcity of pinnipeds in the Mediterranean make them only an occasional prey of Mediterranean white sharks (De Maddalena and Révelart, 2008) (see "Diet," page 188).

In the Mediterranean, there are only two cases of white shark attacks on surfers, one of which is doubtful. This is pretty strange, since surfers are particularly at risk of shark attacks because the area outside the surf zone is a feeding area for sharks. In one case, the victim was a wind surfer. Wind surfers tend to sail well beyond the surf line. In contrast to the Mediterranean, pinnipeds are abundant on the Pacific coast of the U.S. and, correspondingly, white shark attacks on surfers are relatively common (Collier, 2003). Consequently, there is likely a relation between the high presence of pinnipeds and the higher rate of attacks on surfers. However, the number of surfers along the Mediterranean coasts is much lower than along the Pacific coast of the U.S. and the Mediterranean white shark population is without a doubt smaller than the north-eastern Pacific white shark population. Some researchers think that the theory suggesting that great white sharks mistake surf board silhouettes for pinnipeds is not convincing and that most of the time white shark attacks on humans are a form of play or hunting practice. In the areas where surfers tend to aggregate, some white sharks may choose to practice hunting with a surfer as the target because of its resemblance to a pinniped. In this scenario, the attack is not a mistake. However, hypotheses about what may lead a great white shark to bite and release a human without eating him is speculation.

Mediterranean white shark attacks can be categorized by nation as follows: Spain, 1 (doubtful); France, 5 (including 2 doubtful); Italy, 27 (including 7 doubtful); Slovenia, 1; Croatia, 10 (including 1 doubtful); Montenegro, 1; Greece, 4; Turkey, 1 (doubtful); Malta, 4 (including 1 doubtful); Mediterranean (unknown location), 1. The Mediterranean white shark attacks can be categorized by the object of attack as follows: 22 attacks on boats (including 8 doubtful); 17 on bathers (including 2 doubtful); 10 on divers; 2 on surfers (including 1 doubtful); 4 on persons involved in unknown activities. The attacks recorded in the study area span almost 300 years. The most ancient case recorded is a fatal attack that occurred in early June 1721 off the beach called Ponte della Maddalena, in Napoli, Italy (Ricciardi, 1721). The most recent case occurred on October 6, 2008, in Mala Smokova Bay, off Vis (Lissa), Croatia (Branko Dragicevic, pers. comm.; Jakov Dulcic, pers. comm.). Concerning the attack sites, there seems to be no pattern, as the incidents occurred in shallow, deep, warm and cold waters. Great white sharks occasionally come close to shore near populated areas and are encountered more often in zones where the bottom drops off very rapidly. Islands, straits, channels and shoals are also likely attack sites. These large predators patrol those areas because their prey congregates there as well (De Maddalena, 2009) (see "Habitat, Movements and Seasonality," page 184).

There are many reasons why the majority of Mediterranean attacks was recorded along the Italian coasts. Italy has an extended coastline, more than any other Mediterranean nation. The north-south orientation extends well into the Mediterranean, forcing the migratory routes of the white shark's prey to border its coasts. Italy draws the highest number of tourists in the Mediterranean area due to the popularity of these waters for recreation and the large number of historical sites. From a geographic perspective, the Mediterranean areas with the most records of white sharks are the Adriatic Sea, Tyrrhenian Sea and the Channel of Sicily. All three of these areas border Italy. The number of shark attacks in North African nations are underrepresented because they are rarely reported and this kind information never gets wide exposure by the African and European media. In some countries, reports of attacks are intentionally suppressed in order to avoid unwanted reactions from the media that may damage the tourism industry (De Maddalena and Baensch, 2005).

The great white shark attacks by surprise, such that the victim typically never sees the shark until it is too late and it is overwhelmed by the unexpected assault and the violent force with which it is executed. Usually the shark approaches from below, from the side or from behind its victim. Frontal attacks and attacks from above are rare (Burgess and Callahan, 1996; De Maddalena, 2002; Levine, 1996; Miller and Collier, 1980) (see page 192, "Predatory Tactics"). This attack strategy was commonly used in the attacks that occurred in the study area. Unfortunately, the attack modes were not properly described in many cases, which did not allow for a more detailed analysis of this aspect.

Meshing is the most effective method for protecting beaches from great white sharks. Nets are placed parallel to the shore, creating a barrier that prevents the predators from passing through them. Shark nets protect many beaches of Australia and South Africa. However, this method is difficult and extremely expensive to establish and maintain, since nets must be patrolled, cleaned and repaired often. There is also a concern about the effects of meshing on the marine ecosystem (De Maddalena, 2008). The presence of the great white shark in the northeastern Adriatic Sea compelled the local authorities to install antishark nets in order to protect the beaches of the Golfo di Trieste, Istria and the Kvarner (De Maddalena, 2002). Presently, the antishark nets are gone and not used in any Mediterranean country. The present-day lack of white sharks, as well as any other dangerous species of shark, and the fact that attacks have become extremely rare in the entire study area make any kind of protection of local beaches unnecessary. For the same reason, alternative methods for protecting beach enthusiasts, including chemical and electric repellants and protective clothing such as the shark-proof suit of chain mail called Neptunic, are virtually unknown in the Mediterranean.

Fisheries

Official landing statistics of great white sharks in the Mediterranean Sea are nonexistent. In fact, no Mediterranean country has ever reported an official landing statistic for this species. Consequently, this book represents the most complete report on the captures of this species in the Mediterranean fisheries. In most Mediterranean countries, fishermen have always fished for sharks and their meat has always been used for human consumption. However, shark captures have dramatically increased in the past 40 years. As the fisheries for bony fish have been depleted, fishermen have compensated by increasing shark captures (De Maddalena, 2008). An estimated 50 percent of the world shark catch is believed to be taken accidentally during fishing for other species such as tuna (family Scombridae) and swordfish *Xiphias gladius* (Linnaeus, 1758). This unplanned capture of marine animals is called bycatch. The great white shark is, as in the past, taken as a bycatch in many fisheries. In the Mediterranean, there was a fishery of which the white shark was the target species, particularly in the eastern Adriatic Sea and in the Stretto di Messina. However, the importance of the great white shark as a fisheries species is limited because of its low occurrence throughout its range.

In the Mediterranean, the great white shark has been caught by using many different types of fishing gear. The total number of captures recorded in this study is 404 white sharks. These captures are categorized by the type of fishing gear used for capture as follows: 250 (including 72 doubtful) with unspecified fishing gear, 27 with nets of unspecified kind, 72 in tuna traps (including 1 doubtful), 4 with surrounding net (purse seine), 11 with gill nets, 5 with trawls, 6 with longlines, 14 with hook and line (hand line), 1 with pot for prawns (tangled with a line), 10 with harpoons, 2 with a lasso loop, one with a gun and one in a tuna cage (see page 208, "Interactions with Tuna Farming"). Of the 11 caught with gill nets, they are categorized as follows: 4 with a trammel net, 3 with drift nets (2 were *spadare* and 1 was a *thonaille*), 1 with fixed bottom gill net (*palamitara*) and 3 with unspecified gillnets. The 6 caught with longlines are categorized as 1 on bottom longline, 1 on drifting longline and 4 on an unspecified longline.

In the Mediterranean, the tuna migrate from the Atlantic Ocean into the Mediterranean Sea to reach the spawning grounds and then migrate from the Mediterranean to the Atlantic after spawning (see "Habitat, Movements and Seasonality," page 184). A large number of sharks were caught in tuna traps. Some tuna traps operate as the tuna are migrating to the tuna spawning grounds, and other tuna traps operate after the tuna have spawned and are heading back to the Atlantic. In some cases, a tuna trap can be used for both migrations by reconfiguring the trap. The tuna trap consists of a long net, called coast (*costa* in Italian) or tail (*coda*) that is set perpendicular to the shore. The schools of tuna are diverted towards the island (*isola*) or nets forming a pen divided into sections called chambers (*camere*). These chambers are separated by doors (*porte*) that are closed behind the tuna as soon as they enter the chambers. As the tuna run through the corridor and go through the chambers, they reach the last chamber, called the chamber of the death (*camera della morte*). The chamber of death has a movable bottom that is raised, crowding the captured tuna. Then the killing of the tuna (*mattanza*) begins.

In the past, there were 122 tuna traps operating along the Italian coasts (79 were located in Sicily, 21 in Sardinia and the others distributed along the peninsula coast). Other traps were present in almost all Mediterranean countries, including France, Spain,

Croatia, Turkey, Libya, Tunisia, Malta and Morocco (Ravazza, 2005). Captures of great white sharks in the tuna traps were relatively frequent and were recorded in the tuna traps located in Italy, Spain, Croatia, Tunisia and Malta. When a great white shark entered a trap, it was seen as a disaster by the fishermen, because the large shark could damage the nets, keep the tuna away from the trap or even scare the tuna in the trap, causing them to break through the nets. To prevent this disaster at a tuna trap in Scopello, Sicily, Italy, a cow shoulder was placed at the trap entry, allowing the great white shark to satisfy its hunger and move away without entering the trap (Baldassare Carollo, pers. comm.; De Maddalena, 2009). Today, tuna traps have almost completely disappeared from the Mediterranean Sea due to pollution, maritime traffic and unregulated fisheries, which have resulted in these tuna changing their migratory routes to farther from the coastline, thus avoiding the traps (Ravazza, 2005). In present-day Italy, just the tuna traps off Isola Piana and Portoscuso in Sardinia, those off Favignana and Bonagia in Sicily, and one off Camogli in Liguria are still working, while the other tuna traps were forced to stop their activity (Ravazza, 2005). Consequently, captures of great white sharks in tuna traps have also declined (De Maddalena, 2009).

In the Mediterranean, white sharks were not caught just as bycatch. A specific fishery existed in the Messina Strait, Italy, in the 19th century and early 20th century, where the swordfish vessels also targeted great white sharks for their meat, capturing them with a specific kind of harpoon (Gamberini, 1917). A harpoon consists of a long shaft with a forked end into which a detachable dart tip is placed. The dart is attached to a strong cord with kill floats. When the fish is harpooned, the dart detaches and the line with the floats is deployed. Swordfish boats are equipped with a long pulpit at the tip of the bow that serves as a platform for the fisherman to plant the dart into the quarry. The harpooned fish drags the floats until it is exhausted, at which point the fishermen dispatches the fish and boats it. Each swordfish fishing vessel that took white sharks was equipped with a specific kind of harpoon for white sharks that had a shaft measuring anywhere from 400 cm to 500 cm. At that time, white sharks were frequently encountered in the Stretto di Messina. The shark was hit from a pulpit that was more solid than the one used for catching swordfish. This pulpit was located toward the stern. To harpoon the shark, the fisherman stood upright at the stern, then, once he had struck the shark, he opened a trapdoor located under his feet and dropped down into a cavity that was waist high. This cavity was necessary so that the fisherman could maintain his balance and position, keeping the line well stretched so that the shark could not turn and sever the line by biting it (Gamberini, 1917).

The great white shark, like many other sharks, is also taken as bycatch in longline fisheries. Pelagic longlines are single-stranded fishing lines 18 to 72 km long, with an average of 1,500 baited hooks placed on droppers off the main line. Pelagic longlines are the most widespread fishing gear used in the open ocean and are widely used in many parts of the world to catch tuna and swordfish. In some areas, the number of sharks caught by longliners reaches 90 percent of the total catch (De Maddalena, 2008).

Top: This estimated 700–800 cm white shark was encountered on September 9, 2002, off Porto San Giorgio (photograph by Glauco Micheli). *Bottom:* This 535 cm female white shark was caught on May 8, 1987, off Favignana, Isole Egadi, Italy (courtesy Gioacchino Cataldo).

The great white shark was rarely recorded from pelagic longline catches in the Mediterranean. This may be a function of white shark rarity in the epipelagic. It could also be a function of gear selectivity, with larger animals breaking the longline gear and thus avoiding capture.

As far as a recreational fishery for white sharks in the Mediterranean, the great white shark was sometimes hooked, but there is no record in the data bank of a white shark caught on recreational fishing equipment. In a case occurred on July 22, 2000, near Isola del Giglio, Italy, sportfishermen hooked an estimated 200 cm white shark, brought it to the surface, gaffed it, and attempted to bring it into the boat; but the shark bent the gaff hook and freed itself back into the water. Recreational fishing equipment capable of landing a white shark has been developed only in the past 100 years and it is no match for a large white shark.

This 551 cm female white shark was caught in March 1968 in the Bosporus Strait, Turkey (courtesy Hakan Kabasakal).

At one time, there was a bounty placed on white sharks in the region of the eastern Adriatic Sea — between the second half of the 19th century and the early 20th century, at the time of the Austro-Hungarian Empire. Between the years 1872 and 1905, because of the threat that great white sharks posed to humans, the Imperial Maritime Austrian government issued three circulars offering a reward of up to 500 florins for every great white shark captured. These circulars also mentioned other shark species but primarily referred to *Carcharodon carcharias*. To obtain the monetary reward, the fishermen had to present their captured sharks to the Natural History Museum of Trieste, Italy, to verify the species identification (De Maddalena, 2000b, 2002). At the State Archives of Trieste, the orders of payment for these rewards are still in existence. Unfortunately, the species for which they were issued was not listed in most cases (De Maddalena, 2002). This eradication measure apparently produced the desired result, since Marchesetti (1884) reported that from April 1872 to July 1882, twenty-one sharks were presented to the Museo di Storia Naturale of Trieste, while Brusina (1888) reported 24 captures from the same area occurring between September 1868 and September 1887.

Interactions with Tuna Farming

Tuna farming is a relatively new industry in which wild tuna are caught and raised to market size in pens. Tow pens are positioned near where the tuna are caught with purse seines. The tuna are off-loaded into the tow pen, which is towed to permanent tethered pens at another location when full. At the growing pens, the tuna are fed a fat-rich food source in order to produce a high-quality final product. Throughout the world, there are several reports of great white sharks being captured in tuna tow cages and in inshore tuna farm cages. Exact numbers are not known because captures are not always reported. According to Malcolm et al. (2001), in Australia there are unsubstantiated reports of up to 10 to 20 captures of white sharks by the tuna farm industry along with multiple interactions each year. In some cases, the tuna farm staff attempted to release white sharks trapped in tuna cages, but it is only recently that successful releases have occurred (Galaz and De Maddalena, 2004). The tuna farm industry has been recently developed in the Mediterranean Sea, with the first tuna farm in that area being established in 1995. Currently, tuna farming in the Mediterranean is practiced in Spain, Italy, Croatia, Turkey, Cyprus, Greece, Tunisia and Libya. Relationships between great white sharks and tuna farms in the Mediterranean have rarely been described, with only two cases of great white sharks trapped in tuna pens reported.

On June 12, 2002, a 50 m-diameter tuna pen belonging to a European tuna farm company was being towed from Libya to Spain. Off Tripoli, Libya, the towing boat stopped to check the tuna cage and an estimated 500 cm white shark broke through the netting at the bottom and entered the tuna cage. The shark swam in circles inside the cage. It was not observed preying or attacking the tuna, but it is unknown if it fed on one or more tuna during the time it remained inside the cage. The tuna farm staff decided to leave the cage alone with the shark inside and ordered the towing boat to continue its travel. Two days later the shark was gone from the tuna cage. It is uncertain when the shark escaped from the cage, but it made a hole in the net forming the vertical wall of the cage. Concerning the behavior of the white shark, it was similar to that described for the 440 cm female shark trapped in an experimental tuna farm of the SARDI off Port Lincoln, South Australia, June 19–24, 2003 (Ian Gordon, pers. comm.; Kate Rodda, pers. comm.). Two sharks entered the cage, making a small hole in the net just large enough for them to enter. In the cage, both sharks appeared relaxed and swam slowly near the inside edge of the cage. In both cases, the captive white sharks did not attack or even stress the tuna in the cage. Having worked for 9 years in the field of Mediterranean tuna farming, biologist Txema Galaz, who witnessed the

An estimated 500 cm female white shark swims in a tuna cage on June 12, 2002, off Tripoli, Libya (photograph by Lorenzo Millan).

white shark enter the tuna pen off Tripoli, reported that this was the only case of a white shark being trapped in a tuna pen in the Mediterranean and said that he had heard of no other cases. Other species of sharks have been recorded entering tuna pens, in particular the blue shark *Prionace glauca* and the shortfin mako *Isurus oxyrinchus* (Galaz and De Maddalena, 2004).

In October 2009, off Mahdia, Tunisia, an incident caused serious injuries to a number of caged tuna, and three days later a 480 cm female great white shark was seen swimming close to the tuna cage. Two days later, on October 17, 2009, it was discovered that the white shark had broken through the netting at the bottom and had entered the tuna cage. The shark started to attack the tuna, especially the weak individuals that were already injured. The divers attempted to release the shark, but one of them hit the animal around the eye with a knife, causing massive blood loss, and the following day the shark was found dead on the cage bottom (Mohamed Dahmouni, pers. comm.).

The lack of great white sharks in the Mediterranean Sea makes the capture of white sharks in tuna cages in this area a rare event. However, further work is required to accurately estimate the number of sharks that may have been trapped in tuna cages. Tuna have always been a primary food source for Mediterranean white sharks, and interactions between these large predators and tuna traps have been described in detail

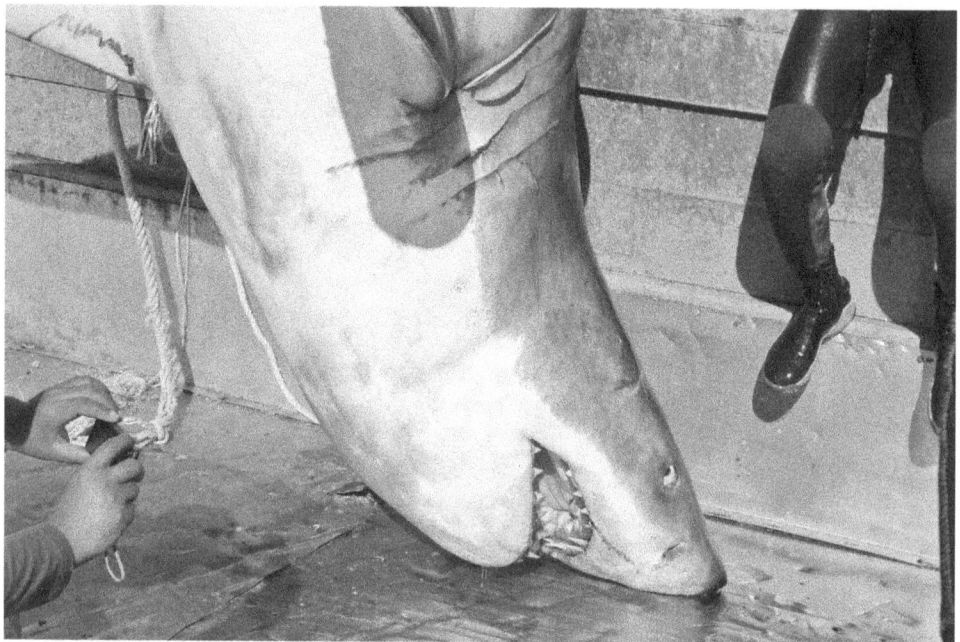

This 480 cm female white shark was caught in a tuna cage on October 17, 2009, off Mahdia, Tunisia (photograph by Mohamed Dahmouni).

(see page 203, "Fisheries"). These tuna pens naturally attract white sharks. Because the number of white sharks in the Mediterranean is considerably fewer today than in the 19th and early 20th centuries, the tuna pens may provide a unique opportunity to study a species that is presently considered very rare in the Mediterranean. As the industry expands, it is expected that other cases will be reported in the near future. However, the lack of reports concerning encounters between white sharks in general from tuna farms in the Mediterranean is additional evidence of the alarming scarcity of these cartilaginous fish in Mediterranean waters.

The great white shark is a protected species in Australia, and the government agency Primary Industries and Resources South Australia (PIRSA) put in place a protocol for the reporting of any interaction between tuna cages and sharks or marine mammals so that appropriate action can be taken. The vulnerable status of white sharks in the Mediterranean Sea and its status as a protected species (see page 216, "Conservation") make necessary the monitoring of interactions between white sharks and the tuna farm industry. When a white shark becomes trapped in a tuna cage, attempts to release it must be made, even though removing a great white shark from a tuna cage is difficult and may present a significant risk to the people involved. Tuna cages should be designed to enable a white shark to escape if it were trapped, possibly through an opening. There should also be a protocol defined for releasing a trapped shark to minimize injury to the shark and the tuna farm personnel. One possible kind of opening may be similar to the one that allowed a successful release of a white shark trapped in a tuna cage off Port Lincoln in June 2003. The gate opening was cut so that it was 9 m deep, then ropes were attached to the following edge at the top and bottom such that when it was pulled it drew the gate open inwards and formed a V-shaped tunnel, expanding the opening so that the shark would have an increased chance of seeing it and run into or sense the net wall (about 3 m long) and turn away from it, through the opening (Ian Gordon, pers. comm.; Kate Rodda, pers. comm.).

Utilization

Roedel and Ripley (1950) erroneously described the great white shark as having no commercial value. The meat of the white shark is as it has been in the past, utilized fresh, frozen, dried-salted, and smoked for human consumption (Compagno, 2001). The meat of this large predator has also been used for livestock feed as fishmeal. Great white shark meat has often been considered of bad quality (Tortonese, 1956). Actually, the meat of this large predator is of excellent quality (Cigala Fulgosi, pers. comm.), to the point that Coles (1919) described it as "the very finest shark, or, in fact, fish of any kind that I have ever eaten, its flavor being quite similar to a big, fat, white shad," and Wood (1959) stated to have "never eaten better fish." Cousins to the great white shark, the shortfin mako *Isurus oxyrinchus* and porbeagle shark *Lamna nasus* are considered excellent to eat and command a high market price. According to the data bank records, many of the large white sharks captured in the past were butchered and sold for human consumption at fish markets.

Even though great white shark meat is of excellent quality for eating, it has an extremely high mercury content (Compagno, 2001). Toxic chemicals that can be absorbed or ingested by animals lower in the food chain are passed up the food chain through consumption. Consequently, top predators like great white sharks concentrate the toxins accumulated from organisms lower in the food chain. This effect is further increased due to the longevity of the great white, since a maximum age estimate of 53 years has been calculated (De Maddalena, 2009).

White shark meat has been sold as smooth-hound (genus *Mustelus*) and swordfish *Xiphias gladius* in European Mediterranean countries (De Maddalena and Baensch, 2005). Because of the stigma associated with white sharks, the meat has rarely been marketed as "white shark meat," although this occurred in Sicily, Italy. Despite the fact that the great white shark is a protected species in the study area, there are many operators in the fields of commercial fisheries, fish trade, and veterinarian fish markets that may have some difficulties distinguishing between one species of the family Lamnidae from another. Determining species identity can be difficult with a whole specimen, and even more if the shark is beheaded or if its jaws have been removed. Therefore, it is likely that occasionally some white shark meat sliced into steaks is sold to the public

A great white shark caught off Favignana, Isole Egadi, Italy, is being butchered (courtesy Gioacchino Cataldo).

at some fish markets in some Mediterranean countries. It would probably be sold as shortfin mako, porbeagle, or under some other name. However, the importance of the trade of white shark meat is very limited because of the low abundance of this species throughout the entire Mediterranean (see page 203, "Fisheries"). Excluding the sharks caught near Majorca, Islas Baleares, Spain, from 1920 to 1976, most of which were sold for human consumption (Morey et al., 2003), only 10 of the 401 sharks captured in the entire Mediterranean were specifically reported to have their meat sold for human consumption.

Whole jaws are collector's items and are often displayed as trophies. Individual teeth are used as souvenirs and as pendants or other jewelry. Dried great white shark jaws and teeth may command high prices on the market, with the jaws and teeth of large sharks over 500 cm-long having the greatest value. The trade of white shark jaws and teeth in the Mediterranean area seems to be almost completely based on sharks caught outside the Mediterranean, from areas such as the waters off Madagascar. There are only two cases recorded in this study where a set of jaws from sharks caught in the Mediterranean were actually put up for sale. It is likely that this has happened on other occasions, but it is also clear that such utilization has been very small and occasional and that most of the time the jaws or teeth were preserved and kept by the fishermen who captured the shark or were given to natural history museums or other scientific institutions (De Maddalena, 2002).

In other parts of the world, there are other uses of white shark parts. Vitamin oil has been extracted from the liver because of its high content of vitamin A. The fins are used in Chinese cooking to prepare the highly desired shark fin soup. The cartilage is promoted as a source of angiogenesis inhibitors for the treatment of cancer, even though there is no solid evidence that this product actually has this kind of property, and the skin has been utilized for leather (Compagno, 2001; Vannuccini, 1999). There is no proof that these products were ever obtained from Mediterranean white sharks.

Great white sharks and their parts have been preserved, cast and put on display in many European museums and other scientific institutions, as will be explored in detail. The utilization of live great white sharks must also be considered. Starting in the late 20th century, the great white shark has been the subject of commercial underwater cage-diving operations in South Australia, South Africa, California and Mexico, but these activities never took place in the Mediterranean. The numbers of white sharks are too few to support a successful cage-diving business. Small great white sharks have been temporarily displayed at a U.S. aquarium. Fortunately, no attempts have ever been made to keep live great white sharks on exhibit in aquariums and oceanariums of the Mediterranean area.

On July 1, 2008, a 125.5 cm newborn white shark was caught in a gill net one km off Altınoluk in the north Aegean Sea. Three days later, on July 4, 2008, another small shark measuring 145 cm TL was caught on a bottom longline in the same place, one km off Altınoluk. The fishermen tried to keep both great white sharks alive for display in a 25-ton marine aquarium. However, both sharks were short-lived, the first dying only 12 hours after capture and the second only 27 hours after capture (Kabasakal and Gedikoğlu, 2008).

Strandings

Strandings of great white sharks have been very rarely reported from the entire Mediterranean area. In fact, only four cases were recorded during this study, and one of them is doubtful. In two of these cases it is specified that the shark was still alive when it was found stranded. All the stranded specimens were large in size. The size of three of them was reported as ranging from 391 cm to over 600 cm.

The earliest reported stranding was in early February of 1839, when a female white shark estimated at over 600 cm in length was either captured or stranded (the various sources differ on this point) in Civitanova Marche, Italy (Bonaparte, 1839; Ernesto Capanna, pers. comm.; Condorelli and Perrando, 1909; De Maddalena, 1998; Metaxà, 1839; Vinciguerra, 1885–1892). In February 1881, a 391 cm white shark was stranded near Beylerbeyi, Turkey (Fergusson, 1996). Between the end of 1908 and the beginning of 1909, after a terrible earthquake took place on December 20, 1908, followed by a huge seaquake, a large white shark was found stranded, still alive, in Maregrosso, Sicily,

Italy (Berdar and Riccobono, 1986; Munthe, 1928). The most recent stranding occurred on November 17, 1992, when a 475 cm male was seen swimming slowly for at least six hours just below the surface in the shallow coastal waters off Tossa de Mar, Spain. After midnight the shark was stranded and moribund on the Playa de la Mar Menuda, where it died one hour later. This shark appeared to be wounded. A cut located at the posterior part of the head, exactly in the cardiac region, is clearly visible in the filmed documentation showing the dying shark. Maybe it was this injury that caused the shark to strand (Barrull, 1993, 1993–1994; Barrull et al., 1999; Barrull and Mate, 2001; Francesc Xavier Viñals Moncusi, pers. comm.).

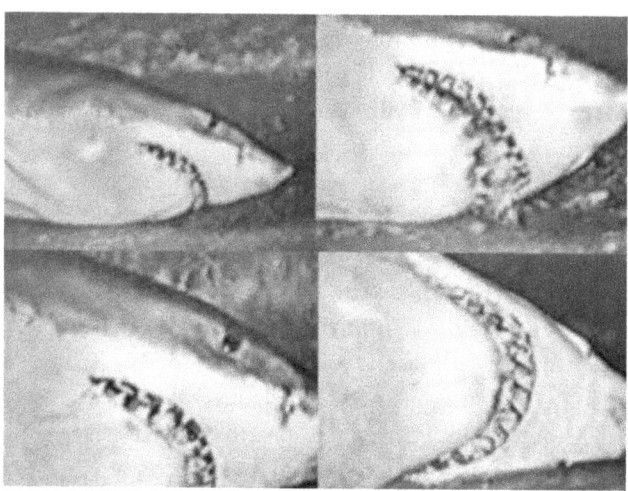

A moribund 475 cm male white shark, stranded on the Playa de la Mar Menuda in Tossa de Mar, Spain, on November 17, 1992 (photograph by Jordi Ferré, courtesy TV Tossa).

Abundance

Based on their limited numbers, white sharks are more vulnerable to overfishing than are bony fish. Despite the fact that few species prey on them, white sharks have long sexual maturation times, low fecundity, long gestation periods, and produce small numbers of young, making them highly vulnerable to over-exploitation (De Maddalena, 2008). Limited evidence indicates that white sharks have few nursery areas, so that even minimal fishing pressure in one of these pupping areas where pregnant females and newborns are concentrated, like the Channel of Sicily, can have devastating results. These fish are unable to withstand extended periods of over-exploitation, and rebuilding white shark populations takes decades.

Great white sharks were, without a doubt, at one time much more abundant in the Mediterranean than they are presently. This is not only true for white sharks, but there is also evidence showing that many species of sharks have suffered a strong decline in the past 40 years, both in the Mediterranean Sea (De Maddalena and Baensch, 2005) and throughout the world (Watts, 2001). Many species of sharks have become uncommon or rare as a consequence of overfishing either the shark or its prey. In addition,

4—Analysis of the Presence of Great White Sharks in the Mediterranean Sea

A 14th century fresco painting portraying a great white shark in the cave of S. Margherita in Castellammare del Golfo, Italy (photograph by Gianfranco Purpura).

humans have an indirect but just as harmful effect on sharks due to depletion of resources, environmental pollution and habitat destruction (De Maddalena and Baensch, 2005).

Unfortunately, a systematic collection of data on captures and landings of white sharks was never performed in the Mediterranean Sea. No governmental institution for fisheries data collection has properly considered the status of this important apex predator in the entire Mediterranean area. Data collected regarding shark captures and landings are scarce and incomplete, with only a few categories considered, including porbeagles *Lamna nasus*, catsharks *Scyliorhinus* sp., smooth-hounds *Mustelus* sp., piked dogfish *Squalus acanthias*, other dogfish sharks (family Squalidae), angelsharks (family Squatinidae), and large sharks (Squaliformes) (Spagnolo, 1999). The great white shark is not even mentioned in this odd listing and would probably fall under the broad category of large sharks. This lack of data prevents any detailed statistical analysis of white shark status in the study area.

The decline of the white shark can be estimated using the data from the data bank. In the ten-year period from 1989 to 1998, there were 85 records from the Mediterranean, while in the following ten years, from 1999 to 2008, there were 46 records. There has been a decrease of 45.88 percent in the number of records. A similar comparison with records dating before 1989 is not possible. The search for data included in the Italian

Great White Shark Data Bank started in 1996 and although much has been done to find all previous records, it is normal that the search for records is more complete from the end of 1980s to the present, while it tends to be less complete going back in time. Another reason that the data collected in recent decades is more complete than that collected farther back in time is the huge growth of media in the last twenty years, along with the mass dissemination of that information. In recent times, there is an increased awareness of the public about the importance of reporting observations of white sharks, in part due to the work done by author De Maddalena as a fundamental part of data collection for the Italian Great White Shark Data Bank.

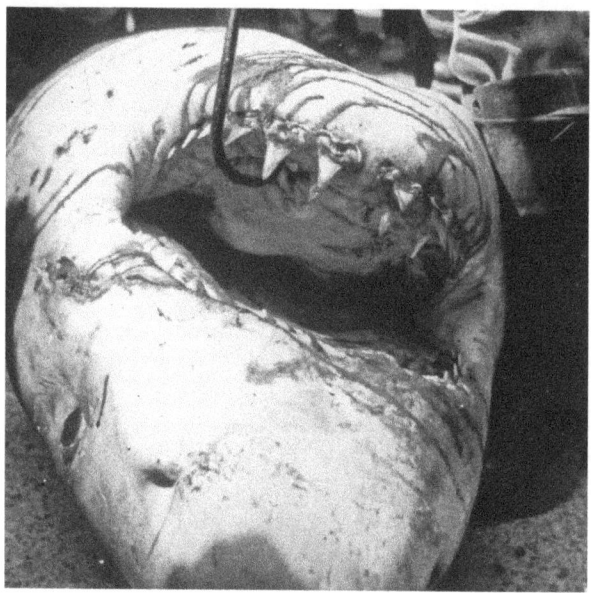

This 540 cm male white shark was caught on April 24, 1980, off Favignana, Isole Egadi, Italy (courtesy Gioacchino Cataldo).

Conservation

In 1996, the great white shark was listed on Annex II (Endangered or Threatened species) of the Protocol concerning Specially Protected Areas and Biological Diversity in the Mediterranean of the Barcelona Convention for the Protection of the Marine Environment and the Coastal Region of the Mediterranean. This should result in full legal protection for the species when the convention is ratified (Wildlife Conservation Society, 2004). In 1999, Italy and Malta implemented this listing by providing national protection for the species, declaring the great white shark a protected species in its territorial waters. However, in Italy, fishermen who unintentionally catch and kill white sharks are not prosecuted. In 2006, the Council of the European Union declared that it shall be prohibited for community vessels to fish for, to retain on board, to tranship and to land the great white shark in all community and non-community waters. This ban applies to all non-community vessels in the waters of the European Community (Council of the European Union, 2006). The basking shark *Cetorhinus maximus* also benefits from the same protection.

Is the great white shark actually protected and safe in the Mediterranean Sea?

Sadly, that is not the case. In the entire Mediterranean area, shark conservation and, more generally, fisheries regulation, are still considered of secondary importance. Surprisingly, no Mediterranean nation invests in white shark research or provides incentives to the few researchers who study the species in this area. There is a lack of interest for what is not an immediate source of profit. The politics that permeate both the academic and the conservation worlds create huge obstacles to the few shark researchers active in the area. The decline in white shark numbers warrants an urgent investigation into the status of the species by all Mediterranean nations. There needs to be effective conservation accompanied by prosecution of those who catch white sharks and proper management of fisheries in which white sharks constitute a bycatch. Besides protecting the white shark, the marine environment where the shark lives and the prey of the white shark also need protection.

The problems facing white sharks are destined to increase with the passing of time and the growth of the human population. The increased human population in the Mediterranean area is the real problem behind the progressive destruction of the marine environment and its creatures. This is rapidly transforming the Mediterranean into a dead sea, a huge desert of water. If human population growth continues, there will be no possible solution to the problem of wildlife destruction. Any conservation measure that may be taken will inevitably fail, being temporary in the best case.

White Shark Materials Preserved in Museums

The study of white sharks preserved in natural history museums of Europe has been a fundamental part of this research program. The results of the study of these items was published recently (De Maddalena, 2006b, 2007; De Maddalena and Zuffa, 2008). With a few recent additions to the items listed in these previous publications, the Italian Great White Shark Data Bank now lists items from 110 great white sharks located in 49 institutions throughout 14 European nations.

Most of these materials are in excellent condition and include many items of great interest. Unfortunately, the data for many of these specimens were lost or are incomplete. Most of these items are skin-mounted specimens or jaws and in some cases they were accompanied by chondrocrania and vertebrae. Other anatomical parts, including single teeth, head, brain, eyes, heart and olfactory bulbs, were rarely preserved. The great white shark items in European museums were preserved using different methods, including taxidermy or skin-mounting, fixation in formalin followed by permanent storage in aqueous solutions of either ethyl or isopropyl alcohols, and drying. In some cases the sharks were not directly preserved, but a model of the shark was cast from a mold of the original specimen.

In the natural history museums of Europe, there are preserved items from 50 white sharks that do not list a capture location. Many of these specimens were likely captured in the Mediterranean. Most specimens with known capture locations are from the Mediterranean (38 specimens on the total of 110 specimens). For the 110 specimens, the number of items preserved by nation is Austria (9), Belgium (2), Bosnia and Herzegovina (1), Croatia (1), Czech Republic (3), Denmark (4), France (23), Germany (6), Italy (43), Monaco (3), Romania (1), Spain (2), Switzerland (2) and United Kingdom (10).

The majority of specimens are very old, with only 12 caught after 1950. The jaws belonging to the Biblioteca Ambrosiana in Milan (without cat. no.), dated from at least 1640–1660, are the oldest white shark material preserved in Europe (De Maddalena, 2006a). The largest specimens with parts preserved in any European museum may be those of which skeletal parts are preserved in the Museo di Anatomia Comparata of Rome, Italy (cat. no. 111-95), and the Museo di Storia Naturale — Sezione di Zoologia ("La Specola") of the University of Florence, in Florence, Italy (cat. no. 6032). Both of these specimens measured approximately 600 cm in length (De Maddalena, 1998). The cast of a 589 cm TOT (565 cm TLn) female great white shark preserved in the Musée cantonal de Zoologie of Lausanne (without cat. no.) is the world's largest white shark that has been reconstructed from a whole specimen (De Maddalena et al., 2003). The largest skin-mounted specimen of Europe is a 522 cm female preserved in the Museo Civico di Storia Naturale of Trieste, Italy (without cat. no.) (De Maddalena, 2006b). The 150 cm female shark preserved in the Senckenberg Forschungsinstitut und Naturmuseum of Frankfurt, Germany (without cat. no.), is the smallest taxidermied white shark preserved in Europe (De Maddalena, 2006b).

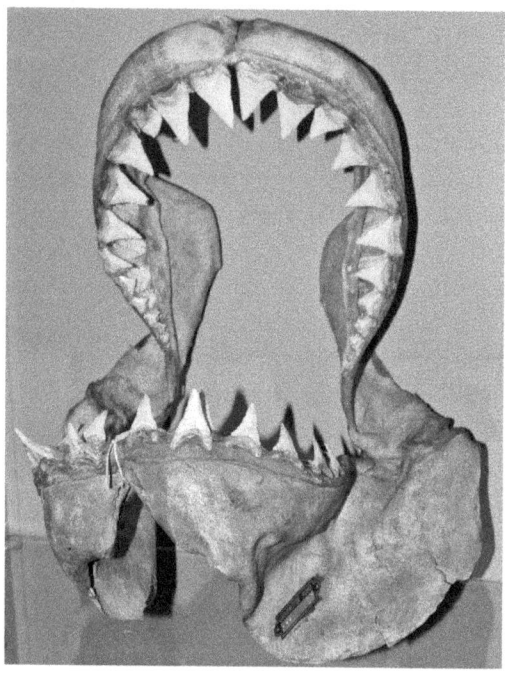

This set of jaws from a white shark caught on an unknown date and in an unknown location, but likely from the Mediterranean, is preserved at the Museo di Anatomia Comparata in Rome, Italy (cat. no. 111-167) (photograph by Ernesto Capanna, courtesy Museo di Anatomia Comparata di Roma, Italy).

Photographic and Filmed Documentation

Of the 596 records of white sharks in the Mediterranean Sea collected in the Italian Great White Shark Data Bank, 122 records are accompanied by photographs. The

4—*Analysis of the Presence of Great White Sharks in the Mediterranean Sea*

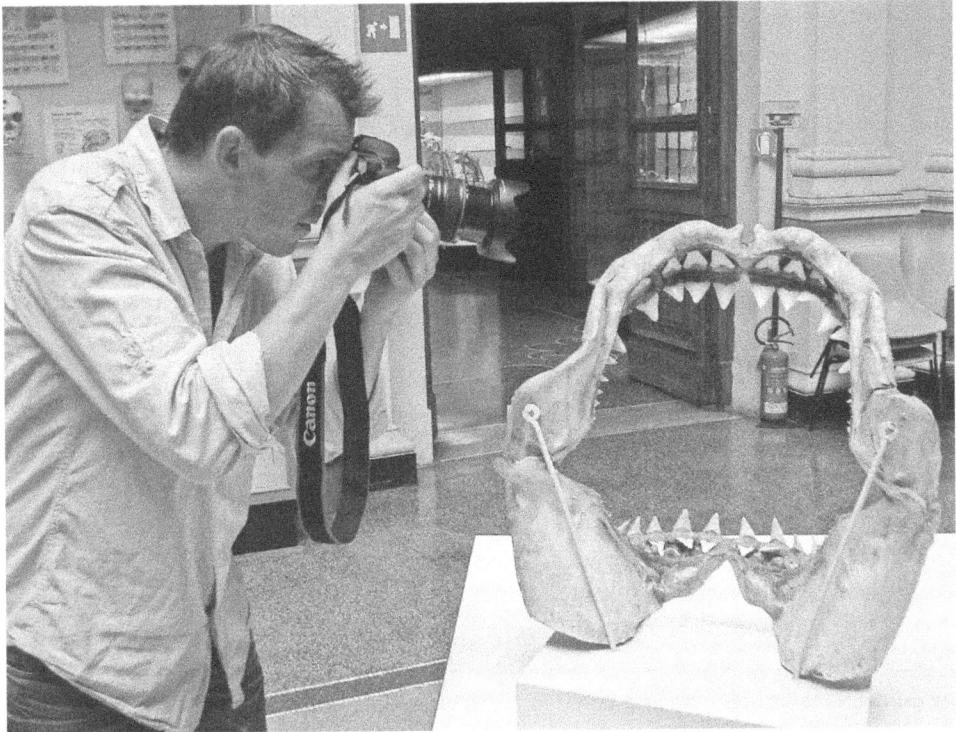

French photographer, Stéphane Granzotto taking photographs of a white shark set of jaws at the Museo Civico di Storia Naturale "Giacomo Doria" of Genova, Italy, for this book (cat. no. C.E. 31916) (photograph by Alessandro De Maddalena, courtesy Museo Civico di Storia Naturale "Giacomo Doria," Genova).

photographic evidence includes pictures portraying live white sharks, dead specimens and anatomical parts. A photo-identification project of live Mediterranean white sharks is currently underway.

While the cases accompanied by photographic evidence are numerous, in only four cases were underwater pictures taken. A female white shark was filmed by cameraman Michel Lobreaux off Favignana, Egadi Islands, Italy, in the late 1960s and the sequence was included in the documentary *Uomini e squali* by Italian director Bruno Vailati (Stefano Carletti, pers. comm.). In July of 1991, an estimated 500 cm female white shark was photographed by Riccardo Andreoli by submerging a camera into the water from the surface on the Pantelleria's shoal off Italy (Riccardo Andreoli, pers. comm.). An estimated 500 cm white shark was photographed off Strombolicchio, Eolie Islands, Italy, in June 1995 or 1996, but unfortunately the picture seems to have been lost (De Maddalena, 2001). On July 12, 2002, an estimated 500 cm specimen, likely a female, was photographed and videographed by Lorenzo Millan by submerging a camera into the water from the surface while the shark swam in a tuna pen off Tripoli, Libya (Galaz and De Maddalena, 2004).

Concerning the video documentation of live great white sharks, in addition to the two short underwater video clips mentioned above that were filmed off Favignana and Tripoli, there is more footage filmed above the water. This includes an estimated 500–600 cm shark filmed off Senigallia on August 27, 1998 (De Maddalena, 2000b), an estimated 700 cm shark off Giulianova between the end of September and early October 1999 (Barone, 1999), an estimated 400–500 cm shark in the Golfe de Valinco, Corsica, France, in August 2009 (Pierre-Henri Weber, pers. comm.), an estimated over–600 cm shark between Sagone and Toulon, France, in December 2009 (Pierre-Henri Weber, pers. comm.), an estimated 400–430 cm female shark off Capraia, Italy, on June 21, 2011 (Cecilia Volpi, pers. comm.), and a 475 cm male shark dying on the Playa de la Mar Menuda in Tossa de Mar, Spain, on November 17, 1992 (Barrull and Mate, 2001). There may also be footage showing an estimated 600 cm shark filmed off Rimini, Italy, at the end of September 1986 (De Maddalena, 2000b); however, it has been impossible to find this video. The footage filmed by Stefano Catalani of an estimated 500–600 cm white shark scavenging a thresher shark carcass on August 27, 1998, off Senigallia, was later included in the documentary entitled '*Méditerranée, requiem pour les requins,*' by French documentarist and photographer Stéphane Granzotto.

To date, no documentary filmmaker has ever dedicated an entire documentary to the great white shark in the Mediterranean Sea based on underwater filming. Attempts made by different teams in Italian and Croatian waters have failed. In fact, the teams were unable to film a single white shark swimming in its natural habitat. In an era in which technology seems to make everything easy and accessible, the realization of an extensive film documentary of Mediterranean great white sharks is a viable goal for the future. Author De Maddalena is at present exploring the possibilities of making such a project reality.

Reporting Specimens of Great White Sharks from the Mediterranean Sea

Everyone who wishes to communicate with author De Maddalena regarding information on great white shark records from the Mediterranean Sea not represented in this book can contact him at the address below. Any kind of record is of interest, including a sighting from a boat, a surface encounter with bathers, an underwater encounter with divers, capture by fishermen, a predatory event, a scavenging event, an attack on a human, or a museum item. With your collaboration, the data collected will be included in the Italian Great White Shark Data Bank, which is important for the continuation and the growth of this research program. Much of the existing data bank is based on the contributions from those who were kind enough to report an encounter. For this reason, a special form is included. The information that is filled in must be as accurate as possible. However, it is typical that some entries are left blank, because it is very rare that all information is available. All data, even though it may be very little, may be very important. If a white shark is caught during sportfishing or commercial fishing, we strongly recommend that the animal be set free after capture with minimal stress. Remember that white sharks are a protected species in the Mediterranean. For released sharks, it is not essential that all the data be supplied. Some data, such as the length, can be estimated by marking the boat gunwale as the shark is along the side prior to release.

The preferred way to collect the data is described as follows. Total length has to be measured in a straight line, from the tip of the snout to the tip of the upper lobe of the caudal fin, with the caudal fin in the natural position (TLn) or with the caudal fin in the depressed position (TOT) (see page 12, "Methods"). If possible, mark the surface where the shark rests and measure the distance between the marks. If the reported length is just an estimate, this must be specified. If the jaws can be examined, the dried upper jaw perimeter (DUJP) has to be measured and, if possible, the enamel height of the largest upper tooth (UAE1 and UAE2) should be measured (see page 12, "Methods"). The preferred weight is for the whole specimen. If the animal has been gutted or beheaded, this must be noted. If the reported weight is just an estimate, this must be specified. The sex, as already described in the chapter about reproduction, is easily recognized by looking at the ventral side of the animal. The males have claspers, the reproductive organs, which are two cylindrical appendices that originate at the base of the pelvic fins. With regards to conserving anatomical parts, it is best that the jaws are preserved as a complete set. If that is not possible, some of the anterior upper and lower teeth should be preserved. The other anatomical parts and the embryos should be frozen or preserved in alcohol or formalin. Preservation of small tissue portions that are frozen or preserved with alcohol will allow future DNA analysis. Preservation of the largest vertebra is highly desired and will allow age estimation.

As for photos, it is advisable to take at least one side photo of the entire subject. Other photos should include details, such as scars and the dorsal fin, to allow subsequent photo identification. In the photo, try to include a meter measuring stick. If a meter stick is not available, something else of known length that can be used to extract size dimensions from the photograph should be used. If dealing with a very large shark and a measuring tool is not available, a human with noted height in cm would be a very good reference. The photo should be broadside with reference to the animal, and the entire animal should be framed with the size reference next to it. For best results, the photographer should stand a few meters away. Information on very large specimens, pregnant females and newborn specimens are considered of special interest.

The form can be sent by normal or electronic post, together with the photographs, to the following address:

Dott. Alessandro De Maddalena
Italian Great White Shark Data Bank
via L. Ariosto 4, I-20145 Milano, Italy
E-mail: a-demaddalena@tiscali.it

Appendix — Mediterranean Great White Shark Report Form

Basic data:
Kind of record (sighting from boat, surface encounter with bathers, underwater encounter with divers, capture by fishermen, predatory event, scavenging event, attack on human, museum item):
Date of capture:
Time of capture:
Place of capture:
Position (latitude and longitude):
Distance from shore:
Depth of sea:
Depth of capture:
Atmospheric conditions:
Sea conditions:
Total length (in a straight line, from the tip of the snout to the tip of the upper lobe of the caudal fin; specify if TLn or TOT):
Weight (specify if intact or gutted):
Sex:
Stomach contents:
Specimen activity at the time of encounter:
Presence of other animals in the area at time of encounter:
Activity of witnesses at time of encounter:
Photographic evidence (if possible, attach):
Specify if data and photos supplied can be published:
Remarks:
References:

Additional data for pregnant white shark females
Number of embryos:
Total length of each embryo:
Weight of each embryo:
Sex of each embryo:
Remarks:
Additional information for preserved white shark specimens
Species: great white shark *Carcharodon carcharias* (Linnaeus, 1758)

BANCA DATI ITALIANA SQUALO BIANCO

ITALIAN GREAT WHITE SHARK DATA BANK

Appendix

Institution name:
Institution address (street, city, state, ZIP, Country, e-mail):
Date of acquisition:
Collector:
Determiner:
Storage or exhibit location:
Catalogue number:
Type of material:
Preservation method:
Enamel height of the largest upper tooth (UAE1 and UAE2):
Dried upper jaw perimeter (DUJP):
Remarks:

Additional information for attacks:
Object of the attack (human, boat, surf, other):
Activity of the victim at time of the attack:
Outcome (fatal/nonfatal):
Provoked/unprovoked:
Part bitten:
Remarks:

Details of contributor
Name and Surname:
Address:
Telephone:
E-mail:
This report should be sent to:
Dott. Alessandro De Maddalena
Italian Great White Shark Data Bank
via L. Ariosto 4, I-20145 Milano, Italy
E-mail: a-demaddalena@tiscali.it

Bibliography

Abela, J. (1989). "Lo squalo bianco più grande del mondo." *Aqua* (January 1989): 20–21.
Albertarelli, M. (1990). "Spaventatevi ma con calma." *Natura Oggi* (August 1990): 30–41.
Ames, J.A., J.G. Geibel, F.E. Wendell, and C.A. Pattison (1996). "White shark-inflicted wounds of sea otters in California, 1968–1992." In *Great white sharks: The biology of* Carcharodon carcharias. Edited by A.P. Klimley and D.G. Ainley. San Diego: Academic Press.
Anonymous (1640–1660). *Volume a disegni colorati rappresentanti oggetti del Museo Settala ceduto dalla Biblioteca di Brera alla Biblioteca Ambrosiana dietro convenuto compenso come da atti, Il 6 Luglio 1907.*
Anonymous (1833). "Annali civili del Regno delle due Sicilie." Napoli: Tipografia del Real Ministero degli Affari Interni.
Anonymous (1891). "Great white shark in the Mediterranean." *Mediterranean Naturalist* 1, no. 4, 76.
Anonymous (1898). "Tremendously big fish caught at Mellieha." *Daily Malta Chronicle*, May 16, 1898.
Anonymous (1908). "Fanciulla straziata da un pescecane." *La Domenica del Corriere*, August 2, 1908.
Anonymous (1924). "Un mostro dei Mari." *La Domenica del Corriere*, June 15, 1924, p. 11.
Anonymous (1926a). "L'orribile fine di un bagnante a Varazze; A circa 200 metri dalla spiaggia viene assalito e sommerso da un pescecane." *Il Secolo XIX*, July 24, 1926, p. 1.
Anonymous (1926b). "Nuovi particolari sulla spaventosa scena di Varazze." *Il Secolo XIX*, July 25, 1926, p. 1.
Anonymous (1930). "Bosforo, squalo si avventa su imbarcazione leggera con due inglesi a bordo" *La Domenica del Corriere*, May 11, 1930.
Anonymous (1931). "Uhvacen morski pas." *Priroda* 21, no. 2–3, 91.
Anonymous (1934). "Morski psi u Kvarneru." *Priroda* 24, no. 8,: 251.
Anonymous (1935). "Morski psi u nasim vodama." *Priroda* 25, no. 8, 254–255.
Anonymous (1938). "La drammatica cattura di uno squalo gigantesco nei pressi dell'Isola d'Elba." Name of newspaper unknown (August).
Anonymous (1947). "Ribari ulovili morskog psa ljudozdera." *Slobodna Dalmacija*, 712.
Anonymous (1951). "Jos nesto o morskom psu." *Priroda* 38, no. 4, 151.
Anonymous (1954). "No em i sjekirom protiv morskog psa." *Slobodna Dalmacija*.
Anonymous (1956). "Le requin capturé par des pecheurs sétois a été acquis par le Musée de Lausanne." *Midi-Libre*, October 14, 1956, p. 4.
Anonymous (1961). "Uno squalo ha divorato un giovane studente che faceva il bagno ad Abbazia (Opatija)." *Il Piccolo*, November 26, 1961.
Anonymous (1963). Untitled. *Primorske novice*, October 25, 1963.
Anonymous (1965). "Uno squalo tigre catturato dai pescatori nello Stretto." *Gazzetta del Sud* (March 1965).
Anonymous (1967). "Grosso smeriglio preso all'isola del Giglio." *La Nazione Sera*, July 24, 1967.
Anonymous (1974). "Lo squalo di Favignana." *Subacqueo* (August-September 1974): 41.
Anonymous (1976). "La psicosi dello squalo." *Mondo Sommerso* (November-December 1976): 49.
Anonymous (1979). "Hanno attaccato un 'morte bianca.'" *Mondo Sommerso* (June 1979).
Anonymous (1984). "A Roccapina, un requin mangeur d'hommes dans les filets des pêcheurs proprianais." *La Corse*, September 16, 1984.
Anonymous (1986a). "Strappa le sardine dalle mani dei pescatori Willy, squalo bianco gigante in Adriatico." *Corriere della Sera*, September 22, 1986.
Anonymous (1986b). "Deux marins-pompiers attaqués par un requin." *Le Provençal*, December 6, 1986.
Anonymous (1987). "Senza titolo." *In-Nazzjon-Taghna*, April 18, 1987.
Anonymous (1989a). "In Tunisia catturato pescecane di due tonnellate." *Corriere della Sera*, February 8, 1989.
Anonymous (1989b). "É ritornato lo squalo bianco: paura nel canale di Piombino." *La Nazione*, May 18, 1989.

Bibliography

Anonymous (1991a). "Un requin blanc de six mètres dans le chalut." *Midi-Libre*, January 10, 1991.
Anonymous (1991b). "Turizmciler çok korktu!," *Sabah* 2 (March 18, 1991): 1.
Anonymous (1992). "Raro squalo bianco pescato in Adriatico." *Corriere della Sera*, March 19, 1992.
Anonymous (1998). "Capo Vaticano, pesca un pesce spada ma viene 'depredato' da uno squalo." *Il Quotidiano*, September 8, 1998.
Applegate, S.P., and L. Espinosa-Arrubarrena (1996). "The fossil history of *Carcharodon* and its possible ancestor, *Cretolamna*: a study in tooth identification." In *Great white sharks: The biology of* Carcharodon carcharias." Edited by A.P. Klimley and D.G. Ainley. San Diego: Academic Press.
Aricò, A. (1990)."La caccia allo squalo continua." *Gazzetta del Sud*, May 26, 1990.
Arrassich, F. (1994). *Trieste in cento cartoline*. Roma: Edizioni La cartolina.
Asensi, J. M. (1977). *El tiburon*. Barcelona: Thor.
Baldridge, H.D. (1973). *Shark Attack Against Man: A Program of Data Reduction and Analysis*. Sarasota: United States Navy, Office of Naval Research.
———. (1974). *Shark attack*. New York: Berkeley Medallion.
Barac, M. (1901). "Prvi morski pas u ovoj sezoni (Psina ljudozder, *Carcharodon verus*)." *Lov. rib. vjesnik*, 10, no. 5, 58.
Barceló, and D.F. Combis (1868). "Catalogo metodico de los peces que habitan o frecuentan las costas de las Islas Baleares." *Rev. Prog. Cien. Ex., Fisc. Nat.*, 18, no. 3–4, 1–46.
Barone, F. (1999). "Ho filmato lo squalo." Name of newspaper unknown.
Barrull, J. (1993). "Polémica sobre la presencia de tiburones blancos en el mar Catalán." *Quercus* (May 1993): 18–19.
———. (1993–1994). "Cita histórica de tiburón blanco *Carcharodon carcharias* (Linnaeus, 1758), en el Mar Catalán (Mar Mediterráneo), documentada con dientes de la mandíbula." *Miscellània Zoològica* 17, 283–285.
Barrull, J., and I. Mate (2000). "Presència del tauró blanc (*Carcharodon carcharias*) al Maresme. Revisió de la seva presència al Mar Català. *IV Jornades Naturalistes del Maresme* 4, 89–93.
———. (2001). "Presence of the great white shark *Carcharodon carcharias* (Linnaeus,1758) in the Catalonian Sea (NW Mediterranean): review and discussion of records, and notes about its ecology." *Annales, Series historia naturalis* 11, no. 1, 3–12.
———. (2001). "Tiburón blanco un gran desconocido de la fauna mediterránea." *Quercus* 184, 24–27.
———. (2002). *Tiburones del Mediterráneo*. Arenys de Mar: Llibreria El Set-ciències, 2002.
Barrull, J., I. Mate, and M. Bueno (1999). "Observaciones de tiburones (Chondrichthyes Euselachii) en aguas de Cataluña (Mediterráneo NO), con algunos aspectos generales de su Ecología." *SCIENTIA gerundensis* 24, 127–151.
Beltrame, M. (1983). "Lo squalo in casa nostra." *Natura Oggi* 3 (July 1983): 52–61.
Ben-Tuvia, A. (1971). "Revised list of the Mediterranean fishes of Israel." *Israel Journal of Zoology* 20, 1–39.
Berdar, A., and F. Riccobono (1986). *Le meraviglie dello Stretto di Messina*. Messina: EDAS.
Bertuccelli, S. (1989). "Caccia all'assassino del mare." *La Repubblica*, February 4, 1989, p. 19.
Betti, R. (1971). *Le prede sportive del sub*. Pisa: Nistri-Lischi Editori.
Biagi, V. (1989). "Squali: Ombre sul mare." *Pesca in mare* 4.
———. (1995). *Memorie della "tonnara" di Baratti —1835–1939*. Baratti: Circolo Nautico Pesca Sportiva Baratti.
Bigelow, H.B., and W.C. Schroeder (1948). *Sharks: Fishes of the western North Atlantic*. Part one, *Lancelets, ciclostomes, sharks*. Memoir Sears Foundation for Marine Research. New Haven: Yale University.
Bonanni, F. (1707). *Musaeum Kircherianum sive musaeum a P. Athanasio Kirchero in Collegio Romano Societatis Iesu iam pridem incoeptum. Nuper restitutum, auctum, descriptum & iconibus illustratum*. Roma: typis Georgii Plachi.
Bonaparte, C. (1839). *Iconografia della fauna italica per le quattro classi degli animali vertebrati*. Tomo III, *pesci*. Roma: Tipografia Salviucci.
Bonomi (1898). "Notizie di caccia e di pesca e note zoologiche: Un gran pesce cane." *Bollettino del Naturalista, Siena* 18, no. 11, 134–135.
Bosnjak, D., and L. Lipej (1992–1993). "Morski psi po svetu in pri nas." *Proteus* 55, 4–9.
Bradaï, M.N., B. Saïdi, M. Ghorbel, A. Bouaïn, O. Guélorget and C. Capapé (2002). "Observations sur les requins du golfe de Gabès (Tunisie méridionale, Méditerranée centrale)." *Mésogée* 60, 61–77.

Bibliography

Brian, A. (1906). *Copepodi parassiti dei pesci d'Italia.* Genova: Istituto Sordomuti.
Bruni, M., and M. Würtz (2002). "The Chondrichthyan fish collection of the Oceanographic Museum of Monaco: history and present status." In *Proceedings of the 4th European Elasmobranch Association Meeting, Livorno (Italy).* Edited by M. Vacchi, G. La Mesa, F. Serena and B. Séret. Roma: ICRAM, ARPAT & SFI.
Brunnich, M.T. (1768). *Ichthyologia Massiliensis, sistens piscium descriptiones eorumque apud incolas nomina.* Accedunt Spolia Maris Adriatici. Hafniae et Lipsiae.
Bruno, C. (1980). "Morte bianca a Favignana." *Mondo Sommerso* (July 1980), 74–75, 125.
Brusina, S. (1888). "Morski psi Sredozemnoga i Crljenog mora (Sharks of the Adriatic and the Black Sea)." *Glasnik hrvatskoga naravoslovnoga družtva* 3, 167–230.
Budker, P. (1971). *The life of sharks.* London: Weidenfeld & Nicolson.
Burgess, G.H., and M. Callahan (1996). "Worldwide patterns of white shark attacks on humans." In *Great white sharks: The biology of* Carcharodon carcharias." Edited by A.P. Klimley and D.G. Ainley. San Diego: Academic Press.
Cadenat, J., and J. Blache (1981). *Requins de Mediterranée et d'Atlantique, plus particulièrement de la Côte Occidentale d'Afrique.* Paris: Editions de l'Office de la Recherche Scientifique et Technique Outremer.
Canestrini, G. (1874). *Fauna d'Italia. Parte terza. Pesci.* Milano: Vallardi, Milano.
Capapé, C., A. Chaldi, and R. Prieto (1976). "Les Sélaciens dangereux des côtes tunisiennes." *Arch. Inst. Pasteur Tunis* 53, no. 1–2, 61–106.
Cappelletti, E. (1989a). "Ecco perché non è stato uno squalo." *Aqva* 34, Supplemento "Squali," 4–11.
_____. (1989b). "La tonnara di Camogli." *Aqva* 38, 106–113.
_____. (1990). "Una grande ombra scura sotto la barca." *Aqva* 49, 30–31.
Capuozzo, T. (1989). "Mare mostrum." *Epoca* (August 1989), 30–32.
Cardellini, S. (1987). "Si è rifatto vivo Willy lo squalo bianco della riviera." *Resto del Carlino*, August 25, 1987.
Carey, F.G., J.G. Casey, H.L. Pratt, D. Urquhart, and J.E. McCosker (1985). "Temperature, heat production and heat exchange in lamnid sharks." *Memoirs of the Southern California Academy of Sciences* 9, 92–108.
Carletti, S. (1973). *Squali di casa nostra. Appendice I all'edizione italiana di Lineaweaver III, T.H. e R.H. Backus. The natural history of sharks [Il libro degli squali].* Milano: Mursia.
Carlsson, J., J.R. McDowell, P. Díaz-Jaimes, J.E.L. Carlsson, S.B. Boles, J.R. Gold, and J.E. Graves (2004). "Microsatellite and mitochondrial DNA analyses of Atlantic bluefin tuna (*Thunnus thynnus thynnus*) population structure in the Mediterranean Sea." *Molecolar Ecology* 13, no. 11, 3345–56.
Carus, J.V. (1893). *Prodromus faunae mediterraneae II: Plagiostomi—Selachoidei.* Stuttgart.
Castro, J. (1983). *The sharks of North American waters.* College Station: Texas A&M University Press.
Cazeils, N. (1998). *Monstres marins.* Rennes: Editions Ouest-France.
Celona, A. (2002). "Su due esemplari di squalo bianco, *Carcharodon carcharias* (Linneo, 1758) catturati nello stretto di Messina nel 1913 e nel 1961." *Doriana, suppl. Annali del Museo Civico di Storia Naturale "G. Doria," Genova* 332, no. 7.
Celona, A. (2002). "Due catture di squalo bianco, *Carcharodon carcharias* (Linneo, 1758) avvenute nelle acque di Marzamemi (Sicilia) negli anni 1937 e 1964." *Annales, series historia naturalis* 12, no. 1, 27–30.
Celona, A., N. Donato, and A. De Maddalena (2001). "In relation to the captures of a great white shark *Carcharodon carcharias* (Linnaeus, 1758) and a shortfin mako, *Isurus oxyrinchus* Rafinesque, 1809 in the Messina Strait." *Annales, series historia naturalis* 11, no. 1, 13–16.
Celona, A., A. De Maddalena, and G. Comparetto (2006). "Evidence of a predatory attack on a bottlenose dolphin *Tursiops truncatus* by a great white shark *Carcharodon carcharias* from the Mediterranean Sea." *Annales, Series historia naturalis* 16, no. 2, 159–164.
Centro Studi Cetacei (1989). "Cetacei spiaggiati lungo le coste italiane" Rendiconto, 1988." *Atti della Società Italiana di Scienze Naturali e Museo Civico di Storia Naturale di Milano* 130, no. 21, 269–287.
_____. (1991). "Cetacei spiaggiati lungo le coste italiane, Rendiconto 1989." *Atti della Società Italiana di Scienze Naturali e Museo Civico di Storia Naturale di Milano* 131, no. 27, 413–432.
_____. (1998). "Cetacei spiaggiati lungo le coste italiane, Rendiconto 1987." *Atti della Società Italiana di Scienze Naturali e Museo Civico di Storia Naturale di Milano* 129, no. 4, 411–432.

Chiocca, G. (1990)."Squalo d'epoca." *Aqva* 51, 10–11.
Cigala Fulgosi, F. "Predation (or possible scavenging) by a great white shark on an extinct species of bottlenosed dolphin in the Italian Pliocene." *Tertiary Research* 12, no. 1 (1990), 17–36.
Cliff, G., L.J.V. Compagno, M.J. Smale, R.P. Van Der Elst, and S.P. Wintner (2000). "First records of white sharks, *Carcharodon carcharias*, from Mauritius, Zanzibar, Madagascar and Kenya." *South African Journal of Science* 96, 365–367.
Cliff, G., S.F.J. Dudley, and B. Davis (1989). "Sharks caught in the protective gill nets off Natal, South Africa" and "The great white shark *Carcharodon carcharias* (Linnaeus)." *South African Journal of Marine Sciences* 8, 131–144.
Coles, R.J. (1919). "The large sharks of Cape Lookout, North Carolina: The white shark or maneater, tiger shark and hammerhead." *Copeia* 69, 34–43.
Collier, R. (2003). *Shark attacks of the twentieth century from the Pacific coast of North America*. Chatsworth: Scientia.
Collier, R.S., M. Marks, and R.W. Warner (1996). "White shark attacks on inanimate objects along the Pacific coast of North America." In *Great white sharks: The biology of Carcharodon carcharias* Edited by A.P. Klimley and D.G. Ainley. San Diego: Academic Press.
Compagno, L.J.V. (1984). *FAO Species Catalogue*. Vol. 4, *Sharks of the world: An annotated and illustrated catalogue of sharks species known to date*. Parts 1 and 2. *FAO Fisheries Synopsis* 125, 1–655.
_____. (2001). *Sharks of the world*. Vol. 2, *FAO Species catalogue for fishery purposes* 1, no. 2, 1–269.
Comune di Pizzo Calabro, Gruppo Acquacultura (1991). *Le tonnare di Pizzo Calabro*. Milano: Jaca Book.
Condorelli, M., and G.G. Perrando (1909). "Notizie sul *Carcharodon carcharias* L., catturato nelle acque di Augusta e considerazioni medico-legali sui resti umani trovati nel tubo digerente." *Bollettino della Società Zoologica Italiana*, 164–183.
Conti, S. (1990). "Ecco lo squalo bianco." *Pesca in mare* (March 1990), 72–77.
Coppleson, V.M. (1958). *Shark attack*. Sydney: Angus and Robertson.
Council of the European Union (2006). "Council Regulation (EC) No 1782/2006 of 20 November 2006 amending Regulations (EC) No 51/2006 and (EC) No 2270/2004, as regards fishing opportunities and associated conditions for certain fish stocks." *Official Journal of the European Union*, L 345 EN, 12 August, 2006, 10–23.
Cousteau, J-M., and M. Richards (1992). *Cousteau's great white shark*. New York: Harry N. Abrams.
Cugini, G., and A. De Maddalena (2003). "Sharks captured off Pescara (Italy, western Adriatic Sea)." *Annales, Series historia naturalis* 13, no. 2, 201–208.
Curzi, C. (1987). "Uno Squalo bianco pescato dalla tonnara di Favignana." *Aqva* 17, 18.
Damiani, G. (1911). "Sovra una *Balaenoptera* del novembre 1910 a Marciana Marina (Elba)." *Bollettino della Società Zoologica Italiana*, 49–57.
Damonte, L. (1993). *L'Estaque, mon village, au temps des pite-mouffe*. Marseille: Paul Tacussel éditeur.
Danker, H. (2001)."Squali in Mediterraneo." *Immersione Rapida Mare* 7 (April-May 2001).
De Beaumont, J. (1957). *Rapport des conservateurs pour l'année 1956*. Musées d'historie naturelle de Lausanne.
De Buen, F. (1926). "Catálogo ictiológico del Mediterráneo espanol y de Marruecos." *Resultado de las campañas realizadas por acuerdos internacionales bajo la dirección del Prof. Odón de Buen* 2, 153–161.
De Maddalena, A. (1997). *Osservazioni sulla presenza e distribuzione di Carcharodon carcharias (Linnaeus, 1758) nel Mare Mediterraneo: segnalazioni e reperti museali*. Tesi di Laurea. Milano: Università degli Studi di Milano, Facoltà di Scienze Matematiche, Fisiche e Naturali.
_____. (1998). "Il più grande esemplare italiano di squalo bianco, *Carcharodon carcharias* (Linnaeus, 1758) individuato nei reperti conservati presso il Museo di Anatomia Comparata dell'Università 'La Sapienza' di Roma." *Museologia Scientifica* 15, no. 2, 195–198.
_____. (1999). *Records of the great white shark in the Mediterranean Sea*. Milano: Alessandro De Maddalena, private publication.
_____. (2000a). "Un programma di ricerca sugli squali bianchi delle nostre acque." *Il Pesce* 4 (August 2000), 64–66.
_____. (2000b). "Sui reperti di 28 esemplari di squalo bianco, *Carcharodon carcharias* (Linnaeus, 1758), conservati in musei italiani." *Annali del Museo Civico di Storia Naturale "G. Doria," Genova* 93, 565–605.
_____. (2002). *Lo squalo bianco nei mari d'Italia*. Formello: Ireco.

_____. (2003). "Historical and contemporary presence of the great white shark *Carcharodon carcharias* (Linnaeus, 1758), in the Northern and Central Adriatic Sea." *Annales, Series historia naturalis* 10, no. 1, 3–18.

_____. (2005). "Alla ricerca del grande squalo bianco: una mostra per conoscere il Leviatano." *Il Pesce* (February 2005), 125–127.

_____. (2006a). "The great white shark, *Carcharodon carcharias* (Linnaeus, 1758) of the Settala Museum in Milan." *Bollettino del Museo civico di Storia naturale di Venezia* 57, 149–154.

_____. (2006b). "A catalogue of great white sharks *Carcharodon carcharias* (Linnaeus, 1758) preserved in European museums." *Časopis Národního muzea, Řada přírodovědná* [*Journal of the National Museum, Natural History Series*] 175, no. 3–4, 109–125.

_____. (2007). *Great white sharks preserved in European museums*. Newcastle-upon-Tyne: Cambridge Scholars.

De Maddalena, A., and H. Baensch (2005). *Haie im Mittelmeer*. Stuttgart: Franckh-Kosmos Verlags-GmbH.

De Maddalena, A., O. Glaizot, and G. Oliver (2003). "On the great white shark, *Carcharodon carcharias* (Linnaeus, 1758), preserved in the Museum of Zoology in Lausanne." *Marine Life* 13, no. 1–2, 53–59.

De Maddalena, A., and W. Heim (2009). *Great white sharks in United States museums*. Jefferson, NC: McFarland.

De Maddalena, A., and J.-P. Herber (2002). "Verrez-vous le grand blanc en Méditerranée?" *Plongeurs International* 47 (June 2002), 68–73.

De Maddalena, A., and L. Piscitelli (2000). "Analisi preliminare dei selaci registrati presso il mercato ittico di Milano." *Bollettino del Museo civico di Storia naturale di Venezia* 52 (April-September 2000), 129–145.

De Maddalena, A., A. Preti, and R. Smith (2005). *Mako sharks*. Malabar: Krieger.

De Maddalena, A., and F. Reckel (2003). "Monstermythos im Mittelmeer." *Unterwasser* (April 2003), 76–81.

De Maddalena, A., and A.L. Révelart (2008). *Le grand requin blanc sur les côtes françaises*. Hyères: Turtle Prod Éditions/Média Plongée.

De Maddalena, A., M. Zuffa, L. Lipej, and A. Celona (2001). "An analysis of the photographic evidences of the largest great white sharks, *Carcharodon carcharias* (Linnaeus, 1758), captured in the Mediterranean Sea with considerations about the maximum size of the species." *Annales, Series historia naturalis* 11 no. 2, 193–206.

De Maddalena, A., and M. Zuffa (2008). "Historical and contemporary presence of the great white shark, *Carcharodon carcharias* (Linnaeus, 1758), along the Mediterranean coast of France." *Bollettino del Museo Civico di Storia Naturale di Venezia* 59, 81–94.

De Michele, V., L. Cagnolaro, A. Aimi, and L. Laurencich (1983). *Il Museo di Manfredo Settala nella Milano del XVII secolo*. Museo Civico di Storia Naturale di Milano.

De Sabata, E., M. Meloni, M. Miliani, and S. Nava (1999). "Bianchi di casa nostra." *Pesca in Mare* 3, 92–99.

Devedjian, K. (1926). *La pêche et les pêcheries en Turquie*. Istanbul: Imprimerie de l'Administration de la Dette Publique Ottoman.

Di Milia R. (1981). *I pesci di Sciacca*. Edizione riveduta da Pasquale Marchese. Sciacca: Tipografia Quartana.

Dieuzeide, R., and M. Novella (1953). *Catalogue des Poissons des côtes Algeriennes*. Vol. 1, *Squales, Raies, Chimères*. Alger: Editions Imbert.

Doderlein, P. (1881). *Manuale ittiologico del Mediterraneo*. Parti 1–2. Palermo: Tipografia del Giornale di Sicilia.

Domeier M.L., and N. Nasby-Lucas (2007). "Annual re-sightings of photographically identified white sharks (*Carcharodon carcharias*) at an eastern Pacific aggregation site (Guadalupe Island, Mexico)." *Marine Biology* 150, 970–984.

Doumet, N. (1860). "Catalogue des poisssons recuilllis ou observés à Cette." *Extrait de la Revue et Magasin de Zoologie*, 1–47.

Dujardin, V. (1890). *Souvenirs du Midi par un homme du Nord*. Céret: Imprimerie Lamiot.

Dumeril, A. (1865). *Histoire naturelle des poissons ou ichtyologie générale*. Vol. 1, *Elasmobranches. plagiostomes et holocéphales ou chimères*. Paris: Librairie Encyclopédique de Roret.

Đurović, E., and S. Obratil (1984). "The living world of aquatic and swamy enviroments (fish, birds, mammals and reptiles)." In *Guide to collections of the regional museum of Bosnia and Herzegovina*. Sarajevo: National Museum SRBiH.

Đurović, E., and S. Obratil (1989). "Živi svijez vodenih i močvarnih staništa (ribe ptice, sisari) i gmizavci." In *Vodič kroz zbirke zemaljskog muzeja Bosne i Hercegovine*. Sarajevo: Zemaljski muzej Bosne i Hercegovine.

Economidis, P.S., and M.-L. Bauchot (1976). "Sur une collection de poissons des mers hélleniques (mers Egée et Ionienne) déposée au Muséum National d'Histoire Naturelle. *Bulletin du Muséum National d'Histoire Naturelle de Paris*, Ser.3, 392 (274), 871–903.

Ellis, R. (1983). *The book of sharks*. London: Robert Hale.

Ellis, R., and J.E. McCosker (1991). *Great white shark*. Stanford: Stanford University Press.

Faber, G.L. (1883). *Fisheries of the Adriatic and the fish of thereof.* London: Bernard Quaritch.

Facciolà, L. (1894). "Cattura di un *Carcharodon Rondeletii* M.H. nel Mare di Messina." *Il Naturalista Siciliano* 13, no. 9, 1–3.

Fanelli, R. (2001). "Il bianco in mezzo al mare." *Pesca in Mare* (June 2001).

Fergusson, I.K. (1996). "Distribution and autecology of the white shark in the Eastern North Atlantic Ocean and the Mediterranean Sea." In *Great white sharks: The biology of* Carcharodon carcharias. Edited by A.P. Klimley and D.G. Ainley. San Diego: Academic Press.

_____. (1998a). "More photos uncovered of Mediterranean white sharks." *Mediterranean Shark News*, May 1998. http://www.zoo.co.uk/~z9015043/news_archives.html.

_____. (1998b). "Juvenile great white shark caught off Tunisia news update." *Mediterranean Shark News*, July 22, 1998. http://www.zoo.co.uk/~z9015043/news_archives.html.

_____. (1998c). "Maltese '7 meter' great white was not a world record." *Mediterranean Shark News*, October 26, 1998. http://www.zoo.co.uk/~z9015043/news_archives.html.

Fergusson, I.K., L.J.V. Compagno, and M.A. Marks (2000). "Predation by white sharks *Carcharodon carcharias* (Chondrichthyes: Lamnidae) upon chelonians, with new records from the Mediterranean Sea and a first record of the ocean sunfish *Mola mola* (Osteichthyes: Molidae) as stomach contents." *Environmental Biology of Fishes* 58, 447–453.

Fiaccarini, R. (1999a). "In Adriatico alla ricerca di Willy lo squalo." *Quotidiano.net*, September 27, 1999.

_____. (1999b). "Willy lo squalo bianco assale un peschereccio." *Quotidiano.net*, September 27. 1999.

Fisher, W., M.L. Bauchot, and M. Schneider (1987). *Fiches FAO d'identification des espèces puor les boesoins de la pêche. (Révision 1). Mediterranée et mer Noire. Zone de Pêche 37*. Rome: FAO.

Francis, M.P. (1996). "Observations on a pregnant white shark with a review of reproductive biology." In *Great white sharks: The biology of* Carcharodon carcharias. Edited by A.P. Klimley and D.G. Ainley. San Diego: Academic Press.

Froldi, M. (2007). "Carloforte, uno squalo bianco squarcia le reti della tonnara." *L'Unione Sarda*, May 16, 2007.

Galaz, T., and A. De Maddalena (2004). "On a great white shark, *Carcharodon carcharias* (Linnaeus, 1758), trapped in a tuna cage off Libya, Mediterranean Sea." *Annales, Series historia naturalis* 14, no. 2, 159–164.

Gasperetti, M. (1998). "Avvistato squalo a Piombino." *Corriere della Sera*, December 29, 1998.

Gemmellaro, C. (1864). "Saggio d'ittiologia del Golfo di Catania." *Atti dell'Accademia Gioenia di Scienze Naturali di Catania* 2, no. 19.

Gessner, C. (1560). *Historia animalium*. Zurich: Froschover.

Gianturco, C. (1978). *Lo Squalo*. Torino: Società Editrice Internazionale.

Giglioli, E.H. (1880). "Elenco dei Mammiferi, degli Uccelli e dei Rettili ittiofagi od interessanti per la Pesca, appartenenti alla Fauna italiana, e Catalogo degli Anfibi e dei pesci italiani." In *Esposizione internazionale di Pesca in Berlino 1880. Sezione italiana, Catalogo degli Espositori e delle cose esposte.* Firenze: Stamperia Reale.

Gilioli, S. (1989). "Caccia al killer." *Sette* 7, 50–61.

Giudici, A., and F. Fino (1989). *Squali del Mediterraneo*. Roma: Atlantis.

Goldman, K.J., S.D. Anderson, J.E. McCosker, and A.P. Klimley (1996). "Temperature, swimming depth, and movements of a white shark at the South Farallon Islands, California." In *Great white sharks: The biology of* Carcharodon carcharias. Edited by A.P. Klimley and D.G. Ainley. San Diego: Academic Press.

Gottfried, M.D., L.J.V. Compagno, and S.C. Bowman (1996). "Size and skeletal anatomy of the giant megatooth shark *Carcharodon megalodon.*" In *Great white sharks: The biology of* Carcharodon carcharias. Edited by A.P. Klimley and D.G. Ainley. San Diego: Academic Press.

Graeffe, E. (1886). "Carcharodon rondeleti." In *Uebersicht der Seethierfauna des Golfes von Triest, etc.* Zoologischen Institut der Universität Wien und der Zoologischen Station in Triest, 7, 446.

Graffione, F. (1991). "Uno squalo terrorizza la Liguria." *La Stampa*, July 31, 1991, p. 9.

Granier, J. (1964). "Les euselaciens dans le Golf d'Aigües-Mortes." *Bulletin du Musée Histoire Naturelle Marseille* 24, 33–43.

Graziosi, G. (1999). "Squalo bianco attacca peschereccio." *Corriere della Sera*, September 27, 1999, p. 19.

Gruber, S.H., and J.L. Cohen (1978). "Visual system of the Elasmobranchs: state of the art 1960–1975." In *Sensory biology of sharks, skates and rays*. Edited by E.S. Hodgson and R.F. Mathewson. Arlington: U.S. Office of Naval Research.

Gubili, C., R. Bilgin, E. Kalkan, S.U. Karhan, C.S. Jones, D.W. Sims, H. Kabasakal, A.P. Martin, and L.R. Noble (2010). "Antipodean white sharks on a Mediterranean walkabout? Historical dispersal leads to genetic discontinuity and an endangered anomalous population." *Proceedings of the Royal Society*, B. 10.1098/rspb.2010.1856.

Gulia, N.G. (1890). Title unknown. *Il Naturalista Maltese* 1, no. 20 (June 1890), 11–12.

Heldreich, Th. de (1878). *La Faune de Grèce: Rapport sur les travaux et recherches zoologiques faites en Grèce*. Première Partie, *Animaux Vertébrés*. Athens: Imprimerie de la Philocalie.

Hemingway, L., and J.C. Devlin (1965). "Le feroci tigri degli oceani." In *Meraviglie e misteri del Regno Animale*. Milano: Selezione dal Reader's Digest.

Hirtz, M. (1932). "Morski psi." *Priroda* 22, no. 7–8, 213–221.

Iglésias S.P. (2011). *Chondrichthyans from the north-eastern Atlantic and the Mediterranean: A natural classification based on collection specimens, with DNA barcodes and standardized photographs*. (Plates and text), Provisional version 05, April 1, 2011.

Imarisio, M. (1998). "Non date la caccia allo squalo." *Corriere della Sera*, August 30, 1998, p. 16.

Jannuzzi, R. (1962). "Caccia grossa in Mediterraneo." *Mondo Sommerso* (June 1962), 20–33, 86–87.

Jolic, S. (1988). "Pro drljivci kukavicjeg srca- Morski psi u Jadranu." *Vikend*, (Ljeto, 1988), 7–78.

Kabasakal, H. (2003). "Historical records of the great white shark, *Carcharodon carcharias* (Linnaeus, 1758) (Lamniformes: Lamnidae), from the Sea of Marmara." *Annales, Series historia naturalis* 13, no. 2, 173–180.

_____. (2008). "Two great white sharks, *Carcharodon carcharias* (Linnaeus, 1758) (Chondrichthyes: Lamnidae), caught in turkish waters." *Annales, Series historia naturalis*, 18, no. 1, 11–16.

_____. (2011). "On an old specimen of *Carcharodon carcharias* (Chondrichthyes: Lamnidae) from the bosphoric waters." *Marine Biodiversity Records* 4, e61.

Kabasakal, H., and Gedikoğlu (2008). "Two new-born great white sharks, *Carcharodon carcharias* (Linnaeus, 1758) (Lamniformes; Lamnidae) from Turkish waters of north Aegean Sea." *Acta Adriatica* 49, no. 2, 125–135.

Kabasakal, H., and E. Kabasakal (2004). "Sharks captured by commercial fishing vessels off the coast of Turkey in the Northern Aegean Sea." *Annales, Series historia naturalis* 14, no. 2, 171–180.

Katuric, M. (1893) "Ihtijolosko- erpetoloske biljeske."

Kelleher, D. (...). "Jaws of death." *Weekender*, no. 2.

Konsuloff, S., and P. Drenski (1943). "Die Fischfauna der Aegais." *Ann. Univ. Sofia Fac. Sci.* 39, 293–308.

Kosic, B. (1903). "Ribe dubrovacke." *Knjiga* 155, Jazu, Zagreb, 48.

Kovačič, M.(1998). "Ihtioloska zirka (Cyclostomata, Selachii, Osteichthyes)." In M. Arko-Pijevac, M. Kovacic and D. Crnkovic, eds. *Prirodoslovna istracivanja rieckog podrucja*. Rijeka: Prirodoslovnog muzeja u Rijeci, pp. 685–698.

Langhoffer, A. (1905). "Popis riba." *G.H.N.D.* 16, 148–169.

Lacépède, B.G.É. (1839). *Histoire naturelle de Lacépède comprenant les cétacés, les quadrupèdes ovipares, les serpents et les poissons*. Paris: Furne.

Last, P.R., and J.D. Stevens (1994). *Sharks and rays of Australia*. Melbourne: CSIRO.

Lawley, R. (1881). *Studi comparativi sui pesci fossili coi viventi dei generi* Carcharodon, Oxyrhina e Galeocerdo. Pisa: Nistri.

Levine, M. (1996). "Unprovoked attacks by white sharks off the South African coast." In *Great white*

sharks: The biology of Carcharodon carcharias. Edited by A.P. Klimley and D.G. Ainley. San Diego: Academic Press.
Lineaweaver, T.H., III, and R.H. Backus (1969). *The natural history of sharks*. Philadelphia: J.B. Lippincott.
Lipej, L. (1993–1994). "Še o izolskem belem morskem volku." *Proteus* 56, no. 5–6, 208–209.
Lipej, L., A. De Maddalena, and A. Soldo (2004). *Sharks of the Adriatic Sea*. Knjiznica Annales Majora: Koper.
Lombardo, G. (1960). "E morì di Domenica." *Mondo Sommerso* (December 1960), 10–15.
Lončarić, R. (2009). "Ralje u Jadranu." July 21, 2009. http://www.geografija.hr/clanci/print-verzija/1489/ralje-u-jadranu.htm.
Long, D.J., K.D. Hanni, P. Pyle, J. Roletto, R.E. Jones, and R. Bandar (1996). "White shark predation on four pinniped species in central California waters: geographic and temporal patterns inferred from wounded carcasses." In *Great white sharks: The biology of* Carcharodon carcharias. Edited by A.P. Klimley and D.G. Ainley. San Diego: Academic Press.
Long, D.J., and R.E. Jones (1996). "White shark predation and scavenging on Cetaceans in the Eastern North Pacific Ocean." In *Great white sharks: The biology of* Carcharodon carcharias. Edited by A.P. Klimley and D.G. Ainley. San Diego: Academic Press.
Lozano Rey, L. (1928). *Ictiologia Ibérica (Fauna Ibérica): Peces (Generalidades, Ciclostomos y Elasmobranquios)*. Vol.1. Madrid: Ed. Museo Nacional de Ciencias Naturales.
Maio, N., P.N. Psomadakis, and M. Vacchi. I Condritti del Museo Zoologico dell'Università di Napoli Federico II. Catalogo degli Elasmobranchi Pleurotremata con note storiche (Pisces, Chondrichthyes, Elasmobranchii). *Annali del Museo civico di Storia naturale "G. Doria" Genova* 96 (2005), 453–481.
Malatesta, N. (1987). "Uno squalo bianco di due metri a Santa Teresa di Gallura." *Aqva* 10, 6.
Malcolm, H., B.D. Bruce, and J.D. Stevens (2001). *A review of the biology and status of white sharks in Australian waters*. Hobart: CSIRO Marine Research.
Maltini, R. (1965). "La casa degli squali." *Mondo Sommerso* (January 1965), 75–83.
Marchesetti, C. (1884). "La pesca lungo le coste Orientali dell'Adria." *Atti del Museo Civico di Storia Naturale di Trieste*, 8.
Marinazzo, E. (2001). "Squalo bianco avvistato ad Apani." *Quotidiano di Brindisi*, April 13, 2001, p. 4.
Marini, F. (1989). "Pochi secondi lunghi un'eternità." *Pesca in mare* 4, 78–79.
Marino, N. (1965). "Un pescecane di otto quintali catturato nella 'secca' di Marzamemi." *Corriere di Sicilia*, (January 1965).
Martelli, F. (1989). "Caccia alla morte bianca." *Pesca in mare* 4, 80–83.
Martin, R.A. (2003). *Field Guide to the Great White Shark*. Vancouver: ReefQuest Centre for Shark Research.
McCosker, J.E. (1987). "The white shark, *Carcharodon carcharias*, has a warm stomach." *Copeia* 1, 195–197.
Megalofonou, P., D. Damalas, and C. Yannopoulos (2005). "Composition and abundance of pelagic shark by-catch in the eastern Mediterranean Sea." *Cybium* 29, no. 2, 135–140.
Melegari, G.E. (1973). *Portofino sub: Guida alla conoscenza dei fondali del promontorio di Portofino*. Genova: ERGA.
Merlo, R. (1979). "Lampione è la loro casa." *Mondo Sommerso* (June 1979), 66–72.
Metaxà, L. (1839). "Smisurato pesce del peso di 4000 libre." *Annali della Società Medico-Chirurgica Metaxà*, 35–38.
Miller, D.J., and R.S. Collier (1980). "Shark attacks in California and Oregon, 1926-1979." *California Fish and Game* 67, no. 2, 76–104.
Miniconi, R. (1987). "Requins de Corse." *Courrier du Parc de Corse*. Numéro special (January 1987), 37.
Miniconi, R. (1994). *Les poissons et la pêche en Méditerranée. La Corse*. Ajaccio: Edition Alain Piazzola et La Marge.
Mizzan, L. (1994). "I Leptocardi, Ciclostomi e Selaci delle collezioni del Museo Civico di Storia Naturale di Venezia—1) Leptocardia, Agnatha, Gnathostomata—Chondrichthyes (esclusi Rajiformes)." *Bollettino del Museo Civico di Storia Naturale di Venezia* 45, 123–137.
Mizzi, C. (1994). "Minn ritratt ta' hajti." *It-Torca*, February 27, 1994, p. 9.
Mojetta, A., T. Storai, and M. Zuffa (1997). "Segnalazioni di squalo bianco *Carcharodon carcharias* in acque italiane." *Quaderni della civica: Stazione idrobiologica di Milano* 22, 23–38.

Mollet, H.F., G.M. Cailliet, A.P. Klimley, D.A. Ebert, A.D. Testi, and L.J.V. Compagno (1996). "A review of length validation methods and protocols to measure large white sharks." In *Great white sharks: The biology of* Carcharodon carcharias. Edited by A.P. Klimley and D.G. Ainley. San Diego: Academic Press.

Mollet, H.F., G. Cliff, H.L. Pratt Jr., and J.D. Stevens (2000). "Reproductive biology of the female shortfin mako, *Isurus oxyrinchus* Rafinesque, 1810, with comments on the embryonic development of lamnoids." *Fishery Bulletin* 98, 299–318.

Montefiori, S. (1998). "Squalo all'attacco, paura nelle Marche." *Corriere della Sera*, August 29, 1998, p. 14.

Moreau, E. (1881). *Histoire naturelle des poissons de la France*. Tome Premier. Paris: Masson.

———. (1892). *Manuel d'ichthyologie Française*. Paris: Masson.

Morey, G., M. Martínez, E. Massutí, and J. Moranta (2003). "The occurrence of white sharks, *Carcharodon carcharias*, around the Balearic Islands (western Mediterranean Sea)." *Environmental Biology of Fishes* 68, 425–432.

Morovic, D. (1950). *Prilog bibliografiji Jadranskog ribarstva*. Split: IOR.

———. (1969). "Iz moje ihtioloske bilje nice." *Priroda* 56, no. 8, 230–232.

———. (1973). "Rijetke ribe u Jadranu." *Pomorski zbornik* 11, 367–383.

Müller, J., and F.G.J. Henle (1838). *Systematische beschreibung der plagiostomen*. Berlin: Veit.

Munthe, A. (1928). *The story of San Michele*. London: Butler & Tanner.

Nania, M. (1999). "C'è uno squalo bianco nel mare di Augusta." *La Sicilia*, November 3, 1999, p. 13.

Navoni, M.M. (2000). "L'Ambrosiana e il museo Settala." In A.A.V.V. *Storia dell'Ambrosiana: Il Settecento*. Milano: CARIPLO.

Ninni, E. (1912). *Catalogo dei pesci del Mare Adriatico*. Venezia: Bertotti.

Norman, J.R., and F.C. Fraser (1937). *Giant fishes, whales and dolphins*. London: Putnam.

Parona, C. (1986). "Notizie storiche sopra i grandi Cetacei nei mari italiani ed in particolare sulle quattro balenottere catturate in Liguria nell'autunno 1896." *Atti della Società Italiana di Scienze Naturali e del Museo Civico di Storia Naturale di Milano* 36, 297–373.

Parona, C. (1919). "Il tonno e la sua pesca." *Memorie Comitato Talassografico Italiano* 68, 1–265.

Perez-Arcas, L. (1878). "Nota sobra los peces escuàlidos *Carcharodon carcharias* (*Carcharias lamia*)." *Anales de la Sociedad Espanola de Historia Natural, Actas* 7, no. 2, 9–19.

Perfetti, M. (1989). "Tutte le pinne del 'Mare Nostrum.'" *Il Mare* (September-October 1989), 8–9.

Perosino, S. (1963). *La pesca*. Novara: Istituto Geografico De Agostini.

Perrier, L. (1938). *Une invasion de requins sur les côtes du Languedoc*. Montpellier: Graille et Castelnau.

Perugia, A. (1881). *Elenco dei pesci dell'Adriatico*. Milano: Hoepli.

Peschak, T.P., and M.C. Scholl (2006). *South Africa's great white shark*. Cape Town: Struik.

Piccinno, F., and A. Piccinno (1979). "Cattura di un enorme *Carcharodon* al largo di Gallipoli (Puglia)." *Thalassia Salentina*, December 30, 1979, 89–90.

Pirino, R., and M. Usai (1991). *Guida agli squali del Mediterraneo*. Firenze: EDAI.

Porqueddu, M. (2000). "Due squali bianchi al largo del Giglio: 'Nessun pericolo.'" *Corriere della Sera*, July 30, 2000.

Postel, E. (1958). "Sur la presence de *Carcharodon carcharias* L. 1758 dans les eaux Tunisiennes." *Bulletin du Muséum National d'Histoire Naturelle de Paris*, Ser.2, 30, 342–344.

Preve, M. (1999). "Grandinata in Liguria, jet si schianta in mare." *La Repubblica*, October 25, 1999.

Pyle, P., M.J. Schramm, C. Keiper, and S.D. Anderson (1999). "Predation on a white shark (*Carcharodon carcharias*) by a killer whale (*Orcinus orca*) and a possible case of competitive displacement." *Marine Mammal Science* 15, no. 2, 563–568.

Quignard, J.-P., and C. Capapé (1972). "Complément à la liste commentée des Sélaciens de Tunisie." *Bulletin de l'Institut National Scientifique et Technique d'Océanographie et de Pêche de Salammbô* 2, no. 3, 445–447.

Quignard, J.-P., and A. Raibaut (1993). "Ichtyofaune de la côte languedocienne (golfe du Lion). Modifications faunistiques et démographiques." *Vie Milieu* 43, no. 4, 191–195.

Quignard, J.-P., A. Raibaut, and J.-P. Trilles (1962). "Contribution à la faune ichthyologique sétoise." *Naturalia monspeliensia., série Zoologie* 3, 61–85.

Radovanovic, R. (1965). "Morski pas i covjek." *Morsko ribarstvo*, no. 5–6, 110–111.

Ramis, D. (1988). "Tiburones en Mallorca." *Brisas* 55, 5–12.

Bibliography

Randall, J.E. (1973). "Size of the great white shark (*Carcharodon*)." *Science* 181, no. 4095, 169–170.

———. (1987). "Refutation of lengths of 11.3, 9.0, and 6.4 m. attributed to the white shark, *Carcharodon carcharias*." *California Fish and Game* 73, no. 3, 163–168.

———. (1986). *Sharks of Arabia*. London: Immel.

Ravazza, N. (2005). "Il tonno fatato." Cosedimare, February 14, 2005. www.cosedimare.com.

Rey, J.C., J.A. Caminas, E. Alot, and A. Ramos (1986). "Captures de requins associées à la pêcherie espagnole de palangre en Méditerranée occidentale, 1984, 1985. 1. Aspects halieutiques." *Rapp. P.-V.Reun., Cons. Inst. Explor. Mer Médit*, 32: 240.

Ricciardi, F. (1721). *Distinta relazione del mostruoso pesce preso da' pescatori Napolitani nella spiaggia detta il Ponte della Maddalena il dì 6 Giugno 1721*. Napoli.

Riggio, G. (1984). "Cattura di *Carcharodon Rondeletii*, Mull. Enl. nelle acque di Capo Gallo e di Isola delle Femine." *Il Naturalista Siciliano* 13, 130–133.

Roedel, P.M., and W.E. Ripley (1950). "California sharks and rays." *Fish Bulletin* 75, 1–88.

Roghi, G. (1962). "Una terribile agonia è la vendetta dell'uomo sui pescecani: l'inferno degli squali." *L'Europeo*, 10.

Rondelet, G. (1554). *Libri de piscibus marinis, in quibus verae piscium effigies expressae sunt*. Lyon: Bonhomme.

Rosselli, D. (1990). "Ricordando un pescatore formidabile." *Lo Spicciolo Nuovo* 9, no. 1, 10.

Roule, L. (1912). "Notice sur les Sélaciens conservés dans les collections du Musée Océanographique." *Bulletin de l'Institut Océanographique* 243, 1–36.

Saïdi, B., M.N. Bradaï, A. Bouaïn, O. Guélorget, and C. Capapé (2005). "Capture of a pregnant female white shark, *Carcharodon carcharias* (Lamnidae) in the Gulf of Gabès (southern Tunisia, central Mediterranean) with comments on oophagy in sharks." *Cybium* 29, no. 3, 303–307.

Sassi, A. (1846). *Catalogo dei pesci di Liguria*. Genova: Ferrando.

Scordia, C. (1935). "Per la biologia del tonno (*Thunnus thynnus* [L].). VIII. Osservazioni sui tonni dello Stretto di Messina eseguite nell'annata 1934–35." *Memorie di Biologia Marina e di Oceanografia* 4, no. 2, 3–38.

Séret, B. (1996). "Le grand requin blanc." *Apnea* 7, hors série, 50–60.

Smale, M.J. and P.C. Heemstra (1997). "First record of albinism in the great white shark, *Carcharodon carcharias* (Linnaeus, 1758)." *South African Journal of Science* 93, 243–245.

Smith, J.V.C. (1833). *Natural history of the fishes of Massachusetts, embracing a practical essay on angling*. Boston: Allen & Ticknor.

Soldo, A., and J. Dulcic (2005). "New record of a great white shark, *Carcharodon carcharias* (Lamnidae) from the eastern Adriatic Sea." *Cybium* 29, no. 1, 89–90.

Soldo, A., and I. Jardas (2002)."Large sharks in the Eastern Adriatic." In *Proceedings of the 4th European Elasmobranch Association Meeting, Livorno (Italy), 2000*. Edited by M. Vacchi, G. La Mesa, F. Serena and B. Séret. Roma: ICRAM, ARPAT & SFI.

Soletti, F. (1987). "Egadi, le isole nella storia." *Aqva* 17, 80–97.

Spagnolo, M. (1999). "Sharks in the Mediterranean." In *Shark utilization, marketing and trade*. Edited by S. Vannuccini. *FAO Fisheries Technical Paper* 389, 1–470.

Stenone, N. (1667). "Canis carchariae *dissectum caput et dissectus piscis ex* Canum Genere." Firenze: Elementorum Myologiae Specimen.

Storai, T., A. Mojetta, M. Zuffa, and S. Giuliani (2000). "Nuove segnalazioni di *Carcharodon carcharias* (L.) nel Mediterraneo Centrale." *Atti della Società toscana di Scienze naturali, Memorie*, Ser. B, 107, 139–142.

Strong, W.R., Jr. (1996). "Shape discrimination and visual predatory tactics." In *Great white sharks: The biology of* Carcharodon carcharias. Edited by A.P. Klimley and D.G. Ainley. San Diego: Academic Press.

Tavernari, C. (1976). "Manfredo Settala, collezionista e scienziato milanese del '600." *Istituto e Museo di Storia della Scienza di Firenze, Annali* 1, no. 1, 43–61.

Terzago, P.M., and P.F. Scarabelli (1677). *Museo ò galeria adunata dal sapere, e dallo studio del Sig. Canonico Manfredo Settala nobile Milanese*. Terza edizione. Tortona: Nicolò e Fratelli Viola.

Tortonese, E. (1938). "Revisione degli Squali del Museo Civico di Milano." *Atti della Società Italiana di Scienze Naturali* 77, 1–36.

———. (1956). *Fauna d'Italia*. Vol. 2, *Leptocardia, Ciclostomata, Selachii*. Bologna: Calderini.

———. (1965). *I pesci e i Cetacei del Mar Ligure*. Genova: Mario Bozzi.

Touret, F. (1992). "Requins, alerte rouge en Méditerranée." *Newlook* 108, 68–70.

Travaglini, F. (1986). "Il far west della pesca." *Aqua* 7, 42–57.

Tricas, T.C., and J.E. McCosker (1984). "Predatory behavior of the white shark (*Carcharodon carcharias*) with notes on its biology." *Proceedings of the California Academy of Sciences* 43, no. 14, 221–238.

Trois, F. (1900). *Catalogo delle collezioni d'anatomia comparata del Regio Istituto Veneto di Scienze Lettere ed Arti dalla fondazione (Gennaio 1867 all'Aprile 1900)*. Venezia: Tipografia Carlo Ferrari.

Uchida, S., M. Toda, K. Teshima, and K. Yano (1996). "Pregnant white sharks and full-term embryos from Japan." In *Great white sharks: The biology of* Carcharodon carcharias. Edited by A.P. Klimley and D.G. Ainley. San Diego: Academic Press.

Vacchi, M., and F. Serena (1997). "Squali di notevoli dimensioni nel Mediterraneo centrale." *Quaderni della Civica Stazione Idrobiologica di Milano* 22, 39–45.

Vanni, S. (1992). "Chondrichthyes." In *Cataloghi del Museo di Storia Naturale dell'Universita di Firenze, Sezione di Zoologia "La Specola"* XI. Atti della Societa Toscana di Scienze Naturali, Memorie, Serie B, 99, 85–114.

Vannuccini, S. (1999). "Shark utilization, marketing and trade." *FAO Fisheries Technical Paper* 389, 1–470.

Vinciguerra, D. (1890). *Guida del Museo di Zoologia della R. Università di Roma—Fauna locale—Specie animali della provincia di Roma esistenti nella nuova collezione*. Part 3, *pesci*. Roma: Istituto di Zoologia della Reale Università di Roma.

Watts, S. (2001). *The end of the line?* San Francisco: WildAid.

West, J. (1996). "White shark attacks in Australian waters." In *Great white sharks: The biology of* Carcharodon carcharias. Edited by A.P. Klimley and D.G. Ainley. San Diego: Academic Press.

Wildlife Conservation Society (2004). "White shark *Carcharodon carcharias*: status and management challenges: Conclusions of the Workshop on Great White Shark Conservation Research." New York: Wildlife Conservation Society.

Wood, F.G. (1959). "Man eats maneater." *Mariner, Marineland of Florida*.

Xuereb, P. (1998). "Marine monster's attack off Marsascala." *Sunday Times*, August 2, 1998, 44–45.

Zuffa, M., G. Van Grevelynghe, A. De Maddalena, and T. Storai (2002). "Records of the white shark, *Carcharodon carcharias* (Linnaeus, 1758), from the western Indian Ocean." *South African Journal of Science* 98, no. 7–8, 347–349.

Index

Numbers in ***bold italics*** indicate pages with illustrations.

abundance 214
Adriatic Sea 10, 33, 68–86
Aegean Sea 10, 91, 159, 181, 183, 213
Agamy Beach 92, 160
Agulhas ring 188
Aix-en-Provence 22, 117
Akko 92, 159
Al Iskandariyya 92, 160, 180
albacore 62, 76, 143, 189
Albania 10, 86, 175, 183
albatross 105, 165, 191
albinism 179
Alboran 18, 114, 181, 188
Alboran Sea 10, 18, 114, 181, 188
Alexandria *see* Al Iskandariyya
Alexandroupolis 86, 156
Algeria 10, 105, 106, 166, 181, 182
Alghero 40, ***41***, 42, 127
Alicante 18, 113
Aliki-Kato Axaia 87, 157
Alopias sp. *see* thresher shark
Altınoluk 91, 213
Alvala 84
ampullae of Lorenzini 9
Ancona 73, 148
Andraitx 21, 116
Antibes 22, 30, 118, 121
Aras Dizra ***103***, 105, 165, ***184***
Arno ***48***, 52, 134
Atlantic bluefin tuna *see* bluefin tuna
Atlantic bonito 52, 88, 134, 158, 189
Atlantic mackerel 76, 189
Atlantic Ocean 9, 10, 187, 188, 204
attacks on humans 3, 4, 11–13, 15, 16, 22, 26, 30, 32, 34–37, 39, 42, 44, 45, 47, 50, 51, 54, 55, 58–60, 66, 69–71, 75, 78, 80, 82, 84–88, 93–95, 105, 113, 118, 120, 121, 124–126, 128, 130–135, 137, 140, 142, 146–148, 150, 153–158, 160, 161, 165, 168, 185, 186, 188, 190–193, 198–203, ***200***, 220, 223, 224
Augusta 45, 65, 66, 137, 144, 196
Australia 188, 203, 208, 210, 213
Auxis rochei see bullet tuna

Bagnara 43, 129
Bakar 79, 152
Bakarac 79, 84, 152, 154

Balaenoptera musculus see blue whale
Balaenoptera physalus see fin whale
Baleares (Islas) 16, ***17***, 18–22, 114–117, 212
Balearic Islands 10, 183, 18–22
Balearic Sea 10, 183, 18–22
baleen whale 190
Banco Chitarro 66
Banco del Mezzogiorno 58
Banco di Pantelleria 63, ***66***, 67
Baratti 43, 48, 50–52, 133, 134, 187
Barcelona Convention for the Protection of the Marine Environment and the Coastal Region of the Mediterranean 216
basking shark 27, 43, 107, 179, 216
Beaulieu 31
Belgrade 82
beli morski volk 175
Beylerbeyi 88, 157, 213
Biblioteca Ambrosiana in Milano 108, 109, ***111***, 218
bijela akula 175
Biodola 50, 132
bird 191
bite action 8
Bizerte 99, 163
Black Sea 10
Blue Grotto 96
blue pointer 175
blue shark 34, 96, 162, 189, 209
blue whale 33, 34, 190
bluefin tuna 4, 41, 76, 88, 93, 158, 187–189
Bocche di Bonifacio 31, 39, 40, 126, 127, 187–189
bogue 31
Bologna 69, 71, 145, 146
Bonagìa 59, 141, 205
bony fish 57, 92, 139, 159, 189, 191, 198, 204, 214
Boops boops see bogue
Bos taurus see calf
Bosnia and Herzegovina 10, 82, 84, 86, 155, 175, 183, 218
Bosporus 10, 88–90, 157–159, ***207***
bottlenose dolphin 21, 61, 66, 67, 70, 117, 142, 145, 147, ***189***, 190, 193, ***194***, 227, 228
Bouches de Bonifacio *see* Bocche di Bonifacio

Bova Marina 68, 145
Brac 85, 155
Brindisi 74, 149
British Museum of Natural History in London 112, 174
Budva 86, 156
bullet tuna 67
Büyükada 88, 90, 158, 159

Calavinagra 41
calf 84, 154
Calypso Deep 10
Camargue 23, ***26***, 119
Camogli 35–37, ***38***, 125, 205
Camperia 61, 142
Canale dell'Isola del Giglio 49
Canale di Montecristo 48
Canale di Piombino 43, ***45***, 51, 187
Canale di Sicilia 65, 180, 181, 182, 183, 184, 214
canavar köpekbaligi 175
Canis lupus familiaris see dog
Cannes 22
Caorle 71, 147
Cap Bon 99, 102, 163, 181
Cap Caveaux 30, 121
Cap Croisette 30
Cap Ferrat 31, 121, 185
Capo d'Anzio 50, 132
Capo della Frasca 40
Capo di Feno 31
Capodistria *see* Koper
Capo Gallo 54, 136
Capo Granitola 62, 65, 144
Capo Rasocolmo 58, 140
Capo Santa Croce 55
Capo Spartivento 68
Capo Testa 39, ***39***, 40, ***40***, 126
Capo Torre Finocchio 40, 41, 55, 137, 186
Capo Vado 33, 124
Capo Vaticano 52, 134
Capra hircus see goat
captivity 213
Carangidae 198
Carcharodon 5
cardiac stomach 8
Caretta caretta see loggerhead sea turtle
Carloforte 41, 127
cartilage 8, 213
cartilaginous fish 5, 189, 198, 210
Cassis 22, 118

Index

Castellammare del Golfo **215**
Castellón 16, 17, 113
Castenedolo 109
Catania 54, 55, 135, 136, 137
Cattolica 73, 148
caudal keels 6, 175, 176
caudal peduncle 6, 175, 176
Caussiniere 31
Cavtat 78, 151
Celle Ligure 34, 124
Centro de Recuperación de Animales Marinos Fundación CRAM 16
Centro Nazionale delle Ricerche 70
cestodes 196
cetacean 13, 32, 34, 41, 55, 61, 95, 122, 190, 191, 193, 194, 196
Cetorhinus maximus see basking shark
Channel of Sicily *see* Canale di Sicilia
Chebba 105, 165
Chelonia mydas see green sea turtle
Chondrichthyes 5
chondrocranium 8, 69, 107, 146, 217
Chromis chromis see damselfish
Ciglio di Terra 54
Circeo 45–47, 49, 130, 131, 186, 187
Civitanova Marche 69, **72**, **178**, 213
Civitavecchia 49, 50, 132
coloration 6, 7, 175, 176, 179
Columbretes (Islas) 16, 17, 113
common dentex 31, 40, 67, 76, 126, 189
common dolphin 99, 163, 190
common dolphinfish 74
common remora 198
common two-banded seabream 54
Congreve Channel 95, 96, 161
conservation 216, 217
copepod 43, 196, **197**
Corfu *see* Kerkira
Corsica 10, 22, 30–32, 121, 122, 182, 220
Coryphaena hippurus see common dolphinfish
Council of the European Union 216
courtship 179
Cres 78, 79, 85, 151, 152
Croatia 10, 15, 78–86, 106, 151–156, 167, 175, 181–183, 185, 187, 202, 205, 208, 218, 220
Croatian Museum of Zagreb 78, 151
cusp 7, 8, 176
cusplets 8, 176
Cyclades Islands 86, 156, 195
Cyprus 10, 92, 108, 159, 183, 187, 208

damselfish 31
Dardanelles 10
Delphinus delphis see common dolphin
dental formula 8
Dentex dentex see common dentex
dermal denticle *see* placoid scale
diet 188–192
Diga Cagno 82
Diplodus vulgaris see common two-banded seabream
Distomum continuum 196
distribution 182–184, **183**
dog 33, 57, 123, 139, 191, 215
Dolmabahçe 90, 158
dolphin 16, 20–22, 24, 26, 29, 31, 40, 41, 44, 46, 48, 49, 55, 56, 58–63, 67, 68, 71, 74, 78, 80, 96, 99, 113, 115, 116, 117, 120–122, 129–132, 137, 139–145, 147, 150, 151, 153, 162, 163, 189–194
dried upper jaw perimeter **6**, 14, 221
Dubrovnik 79, 85, 152, 155
Dugi Otok 78, 80, 151, 153

ear 9
Echeneidae 198
Echthrogaleus coleoptratus 43, 196
Edremit Bay **91**, 159, **180**
Egadi (Isole) 56–65, **63**, **64**, **65**, 138–143, **178**, 182, 194, **197**, **206**, **212**, **216**, 219
Egypt 10, 92, 160, 175, 180, 183
Elasmobranchii 5
Elba (Isola d') 43, 44–47, **46**, **47**, 49–52, 129, 130, 132–134, 165, 186
electroreception 9
electroreceptor 9
embryo 59, 92, 93, 99, 102, 104, 141, 159, 160, 163–165, 180, 181, 221, 223
enamel height **6**, 14, 221
endothermy 7
Enfola 44, 45, **46**, 129, 130, 186
Engraulis encrasicolus see European anchovies
Entente Interdépartementale pour la Démoustication 28
Eolie (Isole) 65, 144, 182, 187, 219
Equus caballus see horse
esophagus 8
Estaque 23, 26, **27**, 178, 228
European anchovies 31
European barracuda 31
European pilchard 56, 189
European spiny lobster 40, 126
Euthynnus alletteratus see little tunny
eyes 5

Falconara 76, 150
Fano *see* Othoni
Faro di Piave Vecchia 76, 149, 198
Faro Santa Croce 66

Favazzina 52, 134
Favignana 56, 59–61, 64, **64**, **65**, 138–143, 194, **197**, 205, **206**, **212**, **216**, 219, 220
Felis silvestris catus see cat
Femine (Isola delle) 55, 136
Fenerbahçe Harbor 88
Figuerolles 26, **29**, 120
Filfla 95–97, **97**, 161, 162, **177**, 178, **184**
filmed documentation 219–221
fin whale 33, 34, 43, 48, 124, 129, 131, 190
finprinting 184, 185
fins 5
fisheries 203–208
Florence 33, 35, 42, 43, 44, 111, 112, 123, 124, 128, 171, 218
Florida Museum of Natural History in Gainesville 198
Foça 91, 159
fork length **6**, 13
fossil 5, 42
France 10, 14, 22–32
Friol (Îles du) 26, 121
Frioul 27

Galeocerdo cuvier see tiger shark
Galera (Isola) 62, 143
Gallipoli 67, **68**, 145, 233
Ganzirri 56–58, 61, **62**, 66, 139, 140, 142, 144178, 185, 186
Genova 11, 34–39, 44, 53, 107, 110–112, 124–126, 129, 134, 170, 171, 173, 174, 219
Ghar El Melh 102, 164
Ghazaouet 105, 166
Gibraltar Strait 10, 187, 188
Giglio (Isola del) 49, 53, 123, 132, 135, 207
gill slits 5
Giulianova 75, **75**, 149, 220
Global Shark Attack File 13, 198
glossopetrae 42
goat 58, 60, 140, 141, 191
Gökçeada Island 91, 159
Golfe d'Ajaccio 31, 32, 122
Golfe de Chafarinas 106, 166
Golfe de Gabès 99, 102–104, **102**, 162, 164, 165, 181, 226, 234
Golfe de Valinco 32, 220
Golfo del Leone 40, 127, 183, 187
Golfo di Baratti 43, **48**, 51, 52, 133, 134, 187
Golfo di Cagliari 41
Golfo di Catania 54, 55, 136
Golfo di Genova 35, **37**
Golfo di Oristano 39, 40
Golfo di Taranto 67
Golfo di Trieste 70, 146, 147, 203
Golfo di Venezia 70, **73**, 186, **189**
Gorgona (Isola) 54, 135
Grado 69, 146
Grampus griseus see Risso's dolphin

Index

grand requin blanc 3, 11, 175
grande squalo bianco
Grau-du-Roi 23, 24, 26, 118–120
great white shark 109, 175
greater amberjack 63, 74, 76, 86
Greece 10, 86–88, 156, 157, 179, 182, 183, 185, 186, 195, 202, 208,
green sea turtle 99, 163, 190
grouper 46, 47, 50, 53, 66, 67, 189
Gulf of Lion *see* Golfo del Leone
Gulf of Mexico 187
Gulf of Pirano *77*, 78

habitat 185–189
haifisch 176
Halkidiki 87, 157, 179
harbor porpoise 190
head 5, *6*
heat-retaining system 7
Herceg Novi 86, 156
Hexanchus griseus see sixgill shark
homing 187
horse 22, 118, 191
Hvar 79, 152
hyoid arch 8
hyomandibular 8
hyostylic jaw suspension 8

Ichthyological Research Society 92
identification 184–185
Ika 85, 155
inedible items 192
International Shark Attack File 198
intestinal valve 8
Ionian Sea 10, 33, 67, 68, 145, 182, 183
Israel 10, 92, 159, 182, 183
Istanbul 88, 89, 91, 92
Istituto di Idrobiologia of the University of Messina in Ganzirri 57, 139
Istituto di Zoologia of Genova 39, 44, 107, 111, 171
Isurus 175
Isurus oxyrinchus see shortfin mako
Isurus paucus see longfin mako
Italian Great White Shark Data Bank 1, 3, 4, 10–12, 198, 216–218, 220
Italy 10, 11, 32–78, 79–81, 106–112, 122–150, 153, 167–174, 176–187, 189, 190, 194–197, 202–208, 211–216, 218–221
Izola 78

Jablanac 78, 151
Jabuka 85, 156
jaquetón blanco 175
jaw *6*, 8, 13, 14, *11*, 13, 14, 16, 17, 22–24, *26*, *27*, 32, 33, 35, *35*, *36*, 39, 40, 43–45, *45*, 53–55, 57, 60, 63, 65, 67, 69, 70, *71*, 74, 85, 87, 88, 92, 96, 97, *97*,

105–114, *110*, *111*, 119, 123–127, 129, 134, 136, 137, 139, 142, 144–146, 150, 155, 157, 159, 166–174, *178*, 211, 212, 217–219, *218*, *219*, 221, 224; protrusion 8

Kabataş 90, *90*, 158
kalb 175
Kapidağ 90, 159
Kavallah 87, 156
kelb abjad 175
kelb el b'har 175
Kelibia 99, 162
Keratsini 87, 156
Kerkira (Corfu) 87, 157
killer whale 95, 196
Konao 78, 151
Koninklijk Belgisch Instituut voor Natuurwetenschappen in Brussels 107, 174
Koper 78, 150,
Korcula 79, 152
Kornati 80, 85, 153, 155
Kraljevica 80, 84, 153, 154, 187
KrK 79, 84, 152, 155
Kvarner 79, *81*, 151, 153, 203, 225

La Ciotat 22, 118
La Formica (Isola) 56–58, *63*, *178*
La Galite 99, 162
La Queue 26
La Seyne-sur-Mer 23, *26*, 118
La Spezia 33, 112, 123
Labin 82
Laboratoire d'Ichtyologie de l'Institut des Sciences et Technologies de la Mer de Sfax 104, 165, 181
Lacona 50, 133
lamb 56, 79, 152, 191
Lamna 175–176
Lamna ditropis see salmon shark
Lamna nasus see porbeagle
Lamnidae 5, 175, 176, 211
Lamniformes 5, 7
Lampedusa 58, 64, 66, 67, 140, 143, 144, 181, 190, 193, *194*
Lampione 59, 140
Languedoc 24, 120
lateral line 9
Lausanne 24, 25, *28*, 120, 176, 218
Le Brusc 23, 119
Lebanon 10, 92, 183
lefkos karkarias 175
Levantine Sea 187
Libya 10, 93, 94, 160, 183, 205, 208, 209, 219
Lido di Venezia 70
Ligurian Sea 10, 33–38, 122–125, 181, 182, 183
Linz 70
Lissa *see* Vis
liver 213
Livorno 42, *44*, 51, 52, 111, 128, 134, 135,

loggerhead sea turtle 21, 57, 117, 157, 162, 190
longfin mako 175, 176
Losinj 85, 155
love bites or mating scars 179, 180
Lugano 36
Lukovo 84, 154
Lumbarda 80, 153
Lyon 30, 107, 110, 169

mackerel sharks 5
Madia 105
Maguelone 14, 24, 28, 120, 176
Mahdia *104*, 165, *184*, 209, *210*
Maiden's Tower *89*, 90, 158
Majorca 18–21, *19*, 114–116, 178, 212
Makryalos 87, 157
Mal di Ventre 39, 126
Mala Smokova Bay 86, 156, 185, 202
Malley 24
Malta 10, 42, 94–99, 160–162, 177, 178, 182–184, 202, 205, 216
Malta Centre for Fisheries Sciences 98
manta 21, 116, 189, 198
Maratea 51, 133
Marciana Marina 43, 49, 50, 52
Maregrosso 55, 213
Marina di Salve 68
Marine Biological Station of Sète 24
marine mammal 190–192, 201, 210
marine reptiles 190
marine turtle *see* sea turtle
Marinella di Sarzana 51
Marinella Selinunte 65
Marmaric Sea 183
Marsala 58, 66, 144
Marsaskala 94, 95, 160
Marsaskala Bay 94
Marsaxlokk 96, 98
Marseille 22, 23, 26–28, *27*, 30, 118, 119, 121 127, 27, 28, 109, 119, 121, 178
Martigues 23, *27*, 119
Martinschizza 84, 154
Marzamemi 56, 57, *61*, 138, 139, 186, 227, 232
mating 179
maximum age 8, 211
maximum depth 186
maximum size 12, 13, 176–179
Mazara del Vallo 59, 60, 62, 65, 140–144
Mediterranean monk seal 96, 120, 190
Mediterranean Sea *9*, 9–10
Mediterranean Shark Research Group 3, 12
Méditerranée, requiem pour les requins 220
Medola 80, 153
Melilla 16, 113
Mellieha 95, 160

239

Index

Messina 52, 54–58, 135, 137, 139, 140, 187, 196, 199, 204, 205
Mexico 185, 187, 213, 229
Micra Rock 87, 157, 185, 186
Miniera del Ginepro 46
Minorca 22, 114
Mljet 78, 85, 151, 155
Moines (îlot des) 31, 122
Mola mola see ocean sunfish
molluscs 33, 123, 191
Mon Repo 87, 157
Monachus monachus see Mediterranean monk seal
Monaco 10, 23, **26**, 27, 32, 111, 111, 119, 173, 183, 218
Montargis 29
Montecristo 48, 131
Montenegro 10, 86, 156, 182, 202,
Monterosso al Mare 33, **35**, 124
Morocco 10, 106, 166, 183, 205
Moscenicka Draga 84
mouth 5
movements 185–189
Mugil sp. *see* mullet
mullet 57, 74, 139, 186, 189
Munxar 94, 95, 160
Musée Cantonal de Zoologie of Lausanne 25, 28, 120, 218
Musée des Confluences in Lyon 107, ***110***, 169
Musée Océanographique de Monaco 23, 26, 27, 234
Musée Vert–Musée d'Histoire Naturelle of Mans 108
Musei Civici of Reggio-Emilia 110, 111, 168, 173
Museo Civico di Storia Naturale "Giacomo Doria" of Genova **11**, 35, **36**, **37**, 44, 53, 107, 110, 112, 125, 129, 134, 170, 173, 174, **219**
Museo Civico di Storia Naturale of Trieste 69, 79, **81**, 106, 153, 167, 208, 218
Museo Civico di Storia Naturale of Venezia 70, **71**, 76, 110, 146, 147, 168, 198
Museo Civico di Zoologia in Rome 69, 235
Museo di Anatomia Comparata of Bologna 69, **71**, 145, 146
Museo di Anatomia Comparata of Rome 69, **72**, 110, **178**, 218, **218**
Museo di Anatomia Comparata of Torino 110, 170
Museo di Storia Naturale e del Territorio of Calci 110, 171
Museo di Storia Naturale e della Strumentazione Scientifica of Modena 32–34, **34**, 110, 123, 150, 170
Museo di Storia Naturale of Ferrara 111, 171
Museo di Storia Naturale of Livorno 111, 171
Museo di Storia Naturale of Milano 33, 108, 109, 111, 123, 168
Museo di Storia Naturale of Palermo 33, 54, 55, 112, 136
Museo di Storia Naturale – Sezione di Zoologia "La Specola" in Florence 33, **35**, 111, 123, 124, 218
Museo di Zoologia in Genova 39, 126
Museo Nacional de Ciencias Naturales in Madrid 17, 114
Museo Regionale di Scienze Naturali in Torino 110, 170
Museo Zoologico in Napoli 110, 169,
Museo Zoologico in Padova 68
Museo Zoologico Universitario in Rome 69
Museu de Zoologia in Barcelona 17, **97**
Muséum d'Histoire Naturelle de Nîmes 23, 108, 118
Muséum d'Histoire Naturelle in Grenoble 108, 172
Muséum d'Histoire Naturelle in Nice 108, 173
Muséum d'Histoire Naturelle in Valence 108, 173
Museum Kircherianum 110, 170, 226
museum materials 217, 218
Muséum National d'Histoire Naturelle of Paris 23, **26**, 103, 105, 108, 165, 171, 172, 230, 233
Museum of History and Culture on Gozo Island 96
Muséum Requien in Avignon 108, 173
Mustelus see smooth-hound

Naples *see* Napoli
Napoli 42, **45**, 110, 128, 169, 186, **200**, 202
Natural History Museum of Zagreb 78, 79, 151, 152
Naturhistorische Museum of Vienna 106, **109**, 167
Naucrates ductor see pilot fish
Nemesis lamna 43, 196
New Zealand 188
newborn 6, 7, 8, **45**, **91**, 91, 92, **180**, 181, 182, 213, 214, 221
Nice 22, 23, 30, 108, 117–119, 121, 173
nictitating membrane 5
Niolon 26, **29**, 120
nostrils 5
Novigrad 85, 155
Numana 72, **73**, 148, **186**

Oblada melanura see saddled seabream
ocean sunfish 58

Oceanographic Museum of Monaco 26, **27**, 111
odontocetes 193
Omiš 85, 155
Opatija 84, 155
Opuzen 78, 151
Orcinus orca see killer whale
Osor 78, 151
Othoni 88
Ovis aries see lamb

Pace 54, 136
Pag 79, 84, 152, 154
Pagasitikos Gulf 87
Palavas 24, 26, 120
Palermo 33, 54, 55, 62, 123, 136, 143, 174
Palinurus elephas see European spiny lobster
Paliouri 87, 157, 179
Pantelleria 60, 63, 66, 67, 141–143, 145, 182, 219
parasites
Paris 23, 26, 29, 32, 53, 54, 61, 100, 103, 105, 108, 118, 165, 166, 171, 175, 227, 22, 230
parturition 181
pas ljudozder 175
Peiraus Port 156
Pelagie (Isole) 58, 59, 64, 66, 67, 140, 143, 144, 182, 190, **194**
Pesaro 71–73, 148, 187
Pescara 70, 147
pescecane 175, 176
peshkagen njeringrenes 175
philopatry 185, 186, 188
Phocoena phocoena see harbor porpoises
photographic documentation 218–220
photo-identification 184
Physeter macrocephalus see sperm whale
Piana (Isola) 39, 111, 126, 205
pig 84, 154, 191
pilot fish 49, 52, 54, 76, 77, **197**, 198
pinniped 190, 191, 201, 202
Piombino 43, 45, 48, 51, 52, 128, 129, 133, 134, 187
Pithos 87, 157, 185, 186
Pizzo Calabro 44, 129
placoid scale **6**, 7
Playa de la Mar Menuda 16, 214, **214**, 220
pointe de l'Espiguette 24, 119
Pointe Rouge 27, 28
Ponte della Maddalena 42, 186, **200**, 202
Pontificium Romanum Archigymnasium 69, 110, 169, 170
Populonia 43
porbeagle 43, 62, 86, 175, 176, 211, 215
Porec 79, 85, 151, 156

Index

Port'Ercole 44, 129
Port Lincoln 208, 211
Porto Barricata 72, 148
Porto San Giorgio 76, **76**, **197**, 198, **206**
Portoferraio 43, 129
Portofino 33–37, **34**, **36**, 123–125, 185
Portopalo di Capo Passero 56, 65, 143
Portoscuso 205
precaudal length **6**, 13
precaudal pits 6
predatory tactics 193–195
pregnant female 59, 92, 102, **102**, 104, 141, 160, 164, 165, 180, 181, 188, 214, 221, 223
Preluka 78, 151
Premiá de Mar 16, 113
Primary Industries and Resources South Australia 210
Primosten 84, 154
Prince Islands 89, 90, 158, 159
Prionace glauca see blue shark
Prirodoslovni Muzej in Rijeka 79
Procchio 51, 133
Procida 43, 129
Propriano 31
Provence 22, 24, 120
Puerto de Mazarrón 17
Punta Baro 82
Punta Chiappa 37, 125
Punta Secca 60, 142
Punta Vagno 36
pyloric stomach 8

qarha levana 175
quadrant 8

Rab 78, 79, 151, 152
Ras Fartas 99, 163
Rava **83**, 85, 155
ray 5, 21, 66, 116, 189
rectum 9
red muscle 7
Red Sea 10
remora 198
Remora remora see common remora 198
Rende 50, 132
Reotore 82
reproduction 180–183
rete mirabile 7
Riccione 70, 147
Ricerche Delfini CTS of Lampedusa 67, 144
Rijeka 79, 80, 82, 152, 154, 187
Rimini 71, 72, 148, 187, 220
ring valve 8
Rio Marina 46, **47**, 130
Riomaggiore 37, 125
Risso's dolphin 22, 117, 190
Riva di Traiano 53, 135
Riva Trigoso 35, **36**, 125
Roccapina 31, 122

Rogoznica 80, 85, 153
Rome 69, 72, 110, 146, 169, 170, 178, 218
Rovinj 80, 84, 153, 154
Rungis 29, 30

S. Lucia al Borgo 54
saddled seabream 39
Sagone 32, 122, 220
Sainte-Marguerite (Île) 22
Sala Civica dei Disciplini 109
Salacak 90
Salistra 88, 158
salmon shark 175
Salvore 78, 150
San Vincenzo 44, 130
Santa Caterina di Pittinurri 40
Santa Croce di Trieste 70
Santa Margherita Ligure 33, 35, 122, 125, 215
Santa Teresa di Gallura 40
Sarajevo 82, 84, 155
Sarda sarda see Atlantic bonito
SARDI 208
Sardina pilchardus see European pilchard
Sardinia 10, 38–42, 126, 127, 182, 186
Saronic Gulf 87
Sarzana 37, 51, 65, 134
scavenging 194, **195**, 199
Sciacca 59, 141, 181
Scilla 42, 43, 50, 128, 129, 133
Scoglio Stella 51, 52, 133, 134
Scomber scombrus see Atlantic mackerel
Scopello 56, 60, **61**, 138, 141, 176, 205
Scorpaena sp. *see* scorpionfish
scorpionfish 53, 135, 189
sea lion 193
Sea of Marmara 10, 88, 89, 90, 158, 159
sea otter 193, 225
sea turtle 21, 21, 40, 49, 57, 87, 96, 99, 103, 117, 126, 132, 157, 162–164, 190–192, 198
seagull 31, 40, 48, 51, 127, 131, 191
seal 26, 120, 190, 193, 201
seasonality 186–189
Secca del Faro 45, 186
Secca del Quadro 46, 47, 49, 187
Secca di Capo d'Omo 49, 132
Secca di Marzamemi 57
Secche di Vada 49, 54, 135
Sedef Adasi 88, 158
segregation 182
Selachimorpha 5
Senckenberg Forschungsinstitut und Naturmuseum of Frankfurt am Main 42, **45**, 128, 218
Senigallia 74, **74**, 149, **195**, 220
Senj 78, 79, **81**, 82, 84, 151, 152, 155

Seriola dumerili see greater amberjack
Serraglio Point 8
Sestri Ponente 34
Sète 14, 21, 23, 24, 28, **28**, **29**, 30, 33, 41, 118, 121, 176, 190
Settala Museum 108, 109, 111, 168
sex ratio 182
sexual maturity 179
Sferracavallo 54, 55, 136
shark fin soup 213
Shark Research Institute 1, 13, 199
shortfin mako 16, 35, 80, 99, 163, 175, 176, 179, 189, 209, 211, 212
Sibenik 85, 155
Sicily 10, 33, 54–68, 135–144, 176, 180–185, 196, 203–205, 211, 213, 214
Sidi Daoud 99–101, **100**, **101**, 103, 162, 164, **184**
Siracusa 58, 140
Six-Four-les-Plages 23
sixgill shark 179
skin 7
Skerki Bank 103, 164
Slovenia 10, 77, 78, 86, 150, 175, 182, 202
smell 9
snout 5
Solenzara River 31, 122
sonar 193
South Africa 179, 185, 191, 203, 213
South Australia 208, 210, 213
Spain 10, 16–22, 93, 97, 113–117, 178, 182, 183, 202, 204, 205, 208, 212, 214, 218, 220
sperm whale 21, 33, 41, 116, 150, 190
Sphyraena sphyraena see European barracuda
Spiaggia del Pezzo 42
Split 79, 85, 86, 151
Sprizze 44, 130
squalo bianco 3, 11, 175
squid 53, 58, 135, 191
Stadival 80, 153
State Archives of Trieste 69, 78, 207
Stazione di Biologia Marina of the University of Lecce 67
Stenella coeruleoalba see striped dolphin
stingray 66
stomach 8, 9
Strait of Gibraltar *see* Gibraltar Strait
Strait of Messina *see* Stretto di Messina
Strait of Sicily 10, 180, 183, 184, 203, 214
stranding 213, 214, **214**
Stretto di Messina 52, 54, 55, 57, 58, 135, 187, 204, 205
striped dolphin 21, 22, 60, 116, 117, 141, 190
Strombolicchio 65, 144, 219

Index

Suez Canal 10
Sus domesticus see pig
Susak 80, 154, 187
Sveti Juraj 78, 79, **82**, 84, 151, 152, 155
Sveti Martin 78, 79, 151, 152
Switzerland 24, 28, 36, 120, 176, 218
swordfish 29, 42, 51, 52, 55, 58, 59, 61, 62, 66, 88, 121, 141, 143, 158, 189, 196, 204, 205, 211
Syria 10, 92, 183
Syros 86, 156, 195

Tabarca **17**, 18, 113
Tarifa 16, 113
Tarragona 17, 114
tauró blanc 175
teeth 5, **6**, 8, 13, 14, 17, 23–26, **26**, 29–31, 39, **41**, 42, 43, **45**, 50, 56, 58, 63, 67, 73, 85, 92, 99, 107, 108, 111, 113, 118–120, 129, 156, 159, 160, 173, 175, 176, 191, 192, 199, 212, 217, 221
Termoli 74, 149
Tetouan 106, 166
Tetrarhynchus megacephalus 196
Thasos 87, 156
Thermaïkos Gulf 87, 157
thresher shark 18, 74, 114, 149, 189, **195**, 196, 220
Thunnus alalunga see albacore
Thunnus thynnus see bluefin tuna
Tiboulen 30, 118, 119
tiburón blanco 175
tiger shark 52
Torre delle Stelle 41, 186
Torre S. Giovanni 67, 78
Torre Testa Rossa 76, 149
Tossa de Mar 16, 113, 214, **214**, 220
total length **6**, 12, 13, 176–179
Toulon 23, 24, 32, 119, 120, 122, 220

Toulouse 30
trematods 196
Tremiti (Isole) 74, 149
Trieste 69, **71**, 78, 79, 81, 106, 146, 147, 153, 167, 203, 207, 208, 218
Trikerion Island 87, 157, 185, 186
Tripoli 93, **94**, 160, 208, 209, **209**, 219
tuna 5, 16–21, 22, 26, 35–36, 39–41, 57; farming 204, 208
Tunis 99,
Tunisia 10, 99–105, 162–166, 175, 180–184, 205, 208–210
Turkey 10, 88–92, 108, 157–159, 180, 182, 183, 202, 205, 207, 208, 213, 40, 41, 49, 55,
Tursiops truncatus see bottlenose dolphin
Tuzla 88, **89**, 158
Tyrrhenian Sea 10, 11, 33, 42, 54, 111, 128–135

Ugljan 80, 153
umbilical scar 92, **180**, 181
University of Cagliari 39–41, 127
University of Florence 33, 35, 111, 123, 124, 218
University of Messina 57, 139
University of Milan 3, 11
University of Neuchatel 24
University of Parma 37
University of Rome 69
Uomini e squali 219
Ustrine 78, 79, 151, 152
uterus 104, 181
utilization 212–214

Vada 49, 54, 132, 135
Valencia 17, 113
Valletta 95, 96, 98
Var 119
Varazze 34, 25, 124

velika bijela ajkula 175
velika bijela psina 175
Venice Lagoon 71
verdoni 34
vertebra 5, 7, 8, 13, 14, 42, 62, 69, **72**, 79, 103, 107, 128, 143, 146, 164, 174, 217, 221
vertebral column 7, 8, 42, 62, 69, **72**, 79, 103, 128, 143
Viareggio 33, 35, 123, 124
Vilassar de Mar 17, 113
Villa Marina 66
Villefranche 31
Vinaroz 17, 114
Vis 86, 156, 202
vision 9
vitamin A 213
Volos 87
Vrboska 79, 152

wahsh 175
weißer hai 175
white death 175
white pointer 175
White Shark Trust 185
Wied iz-Zurrieq 95, 96, 161
Wunderkammer 108

Xiphias gladius see swordfish

yolk 104, 181

Zabbar Sanctuary Museum 95
Zarzis 99, 105, 163
Zemaljski muzej Bosne i Hercegovine of Sarajevo **82**, 84, 155
Zembra Island 99, 163
Zlarin 79, 152
Zoologisk Museum of Copenhagen 99, 105

www.ingramcontent.com/pod-product-compliance
Ingram Content Group UK Ltd.
Pitfield, Milton Keynes, MK11 3LW, UK
UKHW050535150426
5217IPUK00026B/1942